# The Mystery of the Ultimatum Game

Kayoko Kobayashi

# The Mystery of the Ultimatum Game

## Why We Are Predictably Irrational

Springer

Kayoko Kobayashi
Department of Economics
Nanzan University
Nagoya, Japan

ISBN 978-981-97-8978-8 ISBN 978-981-97-8979-5 (eBook)
https://doi.org/10.1007/978-981-97-8979-5

Translation from the Japanese language edition: "Saigo tsuchō gēmu no nazo: Shinka shinrigaku kara mita kōdō gēmu riron nyūmon" by Kayoko Kobayashi, © Kayoko Kobayashi 2021. Published by Nippon Hyoron Sha Co., Ltd. All Rights Reserved.



This Springer imprint is published by the registered company Springer Nature Singapore Pte Ltd.
The registered company address is: 152 Beach Road, #21-01/04 Gateway East, Singapore 189721, Singapore

# Contents

# About the Author

**Kayoko Kobayashi** is an associate professor in the Faculty of Economics, at Nanzan University, Nagoya, Japan. Having completed her doctoral studies at the Graduate School of Economics, University of Tokyo, she focuses on behavioral economics, applied game theory, and law and economics, with special interests in human decision-making from an evolutionary perspective and building social systems that promote well-being for all. This book is her first solo English language publication, and the original Japanese edition of it won the Nikkei Prize for Economics Books (the 64th Nikkei-Keizai Tosho Bunka award) in 2021, an accolade awarded to outstanding economics books. This work also received the Takashima Kunio Jiyu Prize Encouragement Award in 2024. She is the mother of three loving children.

# Chapter 1
# Our demystification tool kit

> *[U]nderstanding the nature of mind is fundamental to building viable theories of social institutions and behavior, of economics and political science. Economics dodged the problem for two centuries with its a priori assumptions of human rationality. But those assumptions are no longer fruitful; they must be replaced by a more veridical theory of the human mind.*
> *—Herbert Alexander Simon (1996), "Models of My Life"*

## 1.1 A Most Puzzling Result

### An economics experiment?

Let's start with a little story. Imagine you are walking through a university campus and come across a poster calling for participants in an "Economics Experiment". All you have to do, it says, is play a game.

> **Economics Experiment**
>
> **Participants Wanted!**
>
> Are you willing to take part in an economics experiment? It's easy. Your task is to play a simple game that involves dividing up real-world money. You can take home any money allotted to you. Participation is entirely free of charge.
>
> Willing participants are asked to meet at the lab on the __ floor of the ______ building at __ o'clock on ______.

You might be puzzled. What on Earth is an economics experiment? And in the form of a game at that. But perhaps it will be fun, and they're even offering a (potential) reward. So along you trot to the economics lab on the prescribed date.


K. Kobayashi, *The Mystery of the Ultimatum Game*,
https://doi.org/10.1007/978-981-97-8979-5_1

On the day, you arrive at the venue and are ushered into a room. The space is divided by a large partition screen. Before long, a figure appears. It must be the researcher. She starts by explaining that there is one other person in this room, and that you will participate in the experiment with that person as a pair. Evidently that person is on the other side of the big partition. You have no idea who they are or what they will be doing. The experiment is anonymous, explains the researcher, and so the individual on the other side will be referred to as Player 1, while you are Player 2.

The researcher gives Player 1 $10 and asks that he (you have no idea if it is a "he", but to keep things simple, we'll refer to Player 1 as such, here and throughout the book) split it between the two of you. Player 1 is free to divide the funds in any way, but if Player 2 (that's you) does not agree, says the researcher, the $10 will be forfeited. Hence, if you say "OK", the funds are divided per Player 1's proposed split, but if you say "No!", the researcher will reclaim the entire amount, and neither of you will receive anything (Fig. 1.1).

**Fig. 1.1** The experiment setup

It's an odd little experiment, you think to yourself, but you feel a little pang of excitement as you wonder how much Player 1 might actually allocate to you. And then in walks the researcher with a single one-cent coin. Unbelievable. Player 1 has given you nothing but a single cent. Naturally, Player 1 is in possession of the remaining $9.99. So Player 1 is proposing to keep $9.99 and allocate you one cent (Fig. 1.2).

**Fig. 1.2** A proposal of $0.01

Now it's your turn. When the researcher asks if you will take the one cent will you accept this proposal, or will you reject it with a resounding "No!"?

**Ultimatum games**

You might perhaps wonder what on Earth is going on here. Well, this is what's known as the **ultimatum game**, one of the most commonly deployed games in the study of game theory.

You might also wonder whether the word "game" means the participants are simply playing around. **Game theory** is a branch of economics and a proper field of study in its own right. It's used to analyze interactions among people, which can include, for instance, bargaining between corporations. Entry-level books on the subject abound, from texts for students to popular-press offerings for the general reader and businessperson, and because of this, I'm sure many readers will have heard of something called the prisoner's dilemma. Each individual model studied in game theory is referred to as a game, and the prisoner's dilemma is one such game, perhaps the most famous of all.

The ultimatum game is another of game theory's most fundamental games and, like the prisoner's dilemma, sees a lot of use. It was originally developed as a simple representation of bargaining interactions, such as buyer–seller bargaining.

**To accept or reject? How much to offer?**

So how about you? How will you react to Player 1 offering you a single one-cent piece? Will you accept the proposal? Or will you reject it?

If you accept it, you will receive $0.01, and Player 1 will walk away with $9.99. If you reject it, the researcher will confiscate the whole $10. You and Player 1 will not be entitled to it. You will both receive nothing. So, which will you choose?

My eldest son caught sight of a draft of this book that I had left lying around. His reaction was, "No way, absolutely not! That's unfair!" I was right there at the time and, intrigued, I asked how he would split the $10 if it were up to him. "Well," he began, "I'd split it half-half, of course." So he would take $5 and give his fellow participant $5 also.

Now, consider yourself in the opposite spot. You've just received $10 from the researcher, and so you have now assumed Player 1's position. Just how much are you willing to offer Player 2 sitting on the other side of the partition?

**The most frequently performed experiment**

This experiment has actually been conducted countless times the world over, to the point that "[t]he ultimatum game is claimed to be one of the most frequently performed of all human experiments today" (Poundstone 2010, p. 114).

We know that when this experiment is conducted, most people around the world behave in accord with what my 10-year-old son said. That is, participants in Player 1's position commonly split the funds down the middle, while those in Player 2's position commonly reject miserly proposals. I have no doubt that many of you would take the same course of action.

Yet while these actions may seem perfectly natural, they are completely different from the theoretical "model answer" predicted by traditional economics and game theory. That model answer is as follows. Player 1 should attempt to keep $9.99 while offering Player 2 one cent, just as Player 1 proposed to you in the lab, and Player 2, having received the one-cent proposal, should accept it. The reason why that turns out to be the model answer is, I think, unlikely to give you a sudden "aha!" moment, even after you've read through the explanation I provide here.

**The unconvincing model answer**

As explained, the ultimatum game is one of game theory's most fundamental games. It appears in many game theory texts, and it's one of the topics I cover without fail when lecturing on game theory at university. But whenever I explain the model answer I just described, I am pretty much always met with an incredulous "That can't be right..." from the students.

And this is not because the thinking behind the answer is at all difficult. I will illuminate the details as we proceed through this book, but the ultimatum game is so structurally and conceptually simple as to be worthy of the moniker "child's play". In fact, looking back, I can't remember there ever being someone to whom the reasoning was impenetrable. And yet the reaction is always something like, "Well, I certainly understand the reasoning, but still..."

Incidentally, this sort of thing almost never occurs with other game theory cases. Even with the famous prisoner's dilemma, although people often behave differently from what the theory would predict, some degree of satisfaction is generally evident on the students' face when things are explained. Game theory itself is really quite a fun field of study, so to my great pleasure as an educator, the overwhelming reaction from students in some of my classes is joy and fascination at a newly revealed way of looking at the world.

Yet the ultimatum game, it is fair to say, is pretty much the only one that elicits "that just doesn't feel right" looks on the faces of a lot of students, no matter how much I explain the reasoning. And to make matters worse, in a sense, that reaction from students actually resonates with me. Even though I teach this material, I have always had an odd feeling, even since my student days, that something is somehow amiss here.

### A most puzzling result

You might think there is something fundamentally off about this oddly named ultimatum game if both the people who teach it and the people to whom it is taught are less than convinced by the traditional reasoning around it. Even so, as a model, the ultimatum game is well suited when it comes to thinking about problems to do with human altruism, fairness, and the like, and it thus took the academic world by storm soon after it was conceived.

The game was taken up by such a wide range of academic disciplines that, in my mind, it is no stretch to say that it has become part of basically any and every field that deals with the nature of humans (and similar animals), from economics through to psychology, cultural anthropology, political science, jurisprudence, ethics, animal behavior, cognitive science, biology, neuroscience, and more. A search for "ultimatum game" on the popular scholarly literature site Google Scholar yields well over 30,000 hits. This veritable firestorm of attention is what led to it being dubbed, as I mentioned earlier, the most frequently conducted experiment. In recent years, researchers have even published papers commemorating 30 years of study on the topic (Van Damme et al. 2014; Güth and Kocher 2014). It is of course exceedingly rare for a research topic to prompt "birthday" celebrations like this.

One reason it has attracted so much attention is the fact that, while the game is structurally very simple, virtually no one behaves the way the theory predicts. In a paper about the dictator game, which can be viewed as a variation on the ultimatum game, Franzen and Pointer (2012) say, "One of the most puzzling results in experimental game theory is that subjects share substantial amounts of resources with their recipients in the dictator game." The question of just how one should go about dissecting this ultimatum game puzzle—the fact that it attracts so much attention while still ultimately evoking incredulity—is one of the challenges that led me to think about it myself, and the motivation for this book.

### What lies at the root of people's choices

In the course of thinking about the ultimatum game puzzle, however, I realized that it probably relates to an even deeper problem. Solving the puzzle, it seems, requires one to think about what human rationality really is. This is equivalent to asking what lies at the root of people's choices, and it also bleeds into the grand question of what the human creature really is at its core.

In this book, I hope to take you on an intellectual journey through such problems guided by the ultimatum game puzzle. Our tools of choice are behavioral economics and evolutionary psychology.[1]

---

[1] My desire to ensure this book properly reflects the latest, cutting-edge research has resulted in pages teeming with footnotes. I have also included end-of-chapter supplements to elucidate some of the finer details. These footnotes and supplements, in general, contain a somewhat higher-level discussion of the issues and can quite happily be skipped over, especially on a first reading. But even simply reading only those that interest you will still provide a taste of the most recent thinking and debate taking place in academic circles.

## 1.2 Tools for Probing the Puzzle

### Homo economicus and rationality

Word association time: What springs to mind when you hear the word "economics"? Money? Interest rates? Supply and demand, profit and loss? These sorts of responses, I must admit, always leave me a little disappointed.

To be sure, economics does indeed deal with dollars and cents and interest to be paid. But what always surprises people, when it is explained, is that the primary focus of economics is humans and how they partake in society. At least, that's how I see it. But perhaps some of the blame for people's surprise when this is explained lies at the feet of economics itself.

The term **Homo economicus** (or **economic man**) is the portrayal of humans as rational agents that has for a long time been used in traditional economic studies. Homo economicus is very logical. She knows what she does and does not like, and in all matters, she is consistent in her reasoning and decisions. She is also very clever, she immediately understands any situation she finds herself in, she can perform complicated calculations with ease, and she knows what the most desirable choice for her is and can arrive at that correct answer instantly. Crucially, she only ever considers her own self-interest when making decisions. She pursues her own interest in a **rational** manner, or in other words, she is an "unremittingly self-regarding actor" (Bowles 2012). (The concept of rationality in economics is a little involved, so for the interested reader, I have provided a supplement to explain it at the end of this chapter.)

This ability to make optimal decisions based on infallible powers of computation have earnt her labels such as "god-like" and "rational fool". You might even think that if such an individual really did exist, she would not be all that much fun to be around. To keep things short, though, from here on out I will follow the lead of Nobel laureate Richard Thaler and refer to Homo economicus as the **Econ** (Thaler 2015).

Views on the Econ have changed considerably, even in economics circles, in recent years. In a voluminous tome, Foundations of Human Sociality, Henrich et al. (2004) note that "experimental economists and others have uncovered large and consistent deviations from the predictions of the textbook representation of Homo economicus." Evidently something was amiss with the concept of the Econ, and the economics world took note of this, with behavioral economics playing a significant role in that shift in views.

> Homo economicus, that hyper-rational, self-interested creature that dismal scientists had embraced since the beginning of time, was dead (if he ever really existed). Hallelujah!
>
> —Levitt and Dubner (2009), "SuperFreakonomics"

---

A note about other works referenced in this book: As is the convention, I refer to references by author name and year of publication. The Latin term "et al.", where it appears after an author's name, means "and others". Works referred to as "mimeo" are unpublished draft manuscripts, and those noted as "forthcoming" are slated for publication in due course.

### Behavioral economics

At times called a heterodox school of thought, **behavioral economics** is a relatively new field of economics that has rapidly gained attention in recent years. One shortcoming is that the name by itself does not immediately clarify what it's all about, but *experiments* are a central aspect of the research methodologies used in this field.

You will not see any beakers or test tubes or Bunsen burners in these experiments, however. The test subjects are human, and the experiments are designed to reveal how humans behave in certain situations. Perhaps surprisingly, though, a lot of people do seem puzzled at the notion of conducting practical experiments in, of all things, economics, so let's consider an example.

One experiment I have conducted with my university students is an auction experiment. A typical auction involves multiple buyers indicating the price they would like to pay for some good—which can be whatever you like, a work of art, fine wine, an antique toy—whereupon the highest bidder makes the purchase at their indicated price. My experiments are set up as sealed-bid auctions in which the students write the price they would be willing to pay—their bids—on a piece such that others cannot see it. Before commencing the bidding process, I ask the students to consider a randomly assigned number, which works as follows. Say that a student is randomly assigned the number 108449. Then I would ask that student to consider whether they would be willing to pay $1084.49 or more for the item.

Curiously, people asked to consider a large number make higher bids in the auction than people asked to consider a small number. This behavior, called the **anchoring effect**, produces some curious results that feel quite implausible, but surprisingly, the behavior is repeatedly observed in all sorts of situations.

Beyond the anchoring effect, we also know, for instance, that people will rate the same wine as tasting better when told that it is more expensive, and that, to give an example from Japan, when presented with the *nami* (standard), *jou* (premium), and *tokujou* (special premium) grades of sushi at a sushi bar, people tend to choose the middle grade. The Econ makes choices based on the value, to her personally, of the specific option, so she should not be at all influenced by where the option appears in some sort of ranking, whether at the top or bottom or bang in the middle. She also judges the taste of wine based on the substance itself, so the alleged value of the wine should have no impact on her assessment.

Economic theory has so far focused on figuring out how people would have to, or should, behave assuming they act like an Econ. Economics experiments, meanwhile, look at how humans actually behave when presented with various decision-making scenarios. And through numerous such experiments, behavioral economics has demonstrated that humans do not always act in the rational way that economics would predict.

### Anomalies

These sorts of aberrant behaviors are called "anomalies" in the literature. They are "odd" behaviors, in the sense that they differ from what an Econ would do, exhibited by large numbers of people. All the examples I just gave are anomalies—bidding higher after being asked to consider a large number, tending to choose the middle-ranked option, and feeling that wine tastes better when you're told it's more expensive. They are peculiar actions or perceptions that, while irrational, transcend individual differences and show up time and time again across large numbers of people.

Behavioral economics, through experiments into countless such anomalies, has accumulated an astonishing amount of evidence showing that humans are not Econs, and that divergences from Econ-like behavior are not mere exceptions to the rule or random errors. There are, in fact, systematic traits of such anomalous behavior and perception common across many individuals.

When we use the tools of behavioral economics to think about game theory problems, we call this behavioral game theory. And the ultimatum game that I tackle in this book is one of the most well-known anomalies in the game theory space.

**Very very rational?**

Now, I often see people claim that behavioral economics has shown that humans are irrational because they are not rational beings. This is the sort of conclusion you might be tempted to accept at face value, but this idea is quite vehemently rejected by Daniel Kahneman, recipient of the Nobel Award in Economics and often referred to as one of the fathers of behavioral economics, who has this to say: "I often cringe when my work with Amos is credited with demonstrating that human choices are irrational, when in fact our research only showed that Humans are not well described by the rational agent model." (Kahneman 2011). His point: just because humans do not behave in the way that economics predicts, that does not mean that humans are irrational.

You might be wondering what on Earth this could mean. How can humans not be rational agents and yet not be irrational at the same time?

I think the answer is that humans are actually *very very rational*. Now, you can be forgiven for thinking I'm talking in circles here. How in the name of all that is sacred can humans be not rational nor irrational but somehow very very rational? It all just sounds like wordplay and riddles. Well, as you may have guessed, the biggest theme of this book is how one should go about deciphering this problem.

**Evolutionary psychology**

The other tool we will be using to probe the puzzle is evolutionary psychology. Now, you might wonder what evolution, something normally associated with the field of biology, could possibly have to do with all this. Children would learn in grade school that evolution is what shaped the physical bodies of humans and other animals. In recent years, though, evolutionary psychologists have argued that evolution also shaped the way we think—that is, our minds.

The idea that evolution has shaped not only people's physical bodies but their minds as well may be somewhat unfamiliar, perhaps even difficult to understand, for some. If so, don't fret. In this book, I hope to walk you through what this all means.

### Structure of the book

I've structured this book as follows. Chapter 2 discusses the essence of the ultimatum game, or what it boils down to, and why it is a puzzle. As explained, the ultimatum game involves a sum of money being split between two participants, Player 1 and Player 2, and experiments around the world have shown that both participants behave quite differently from what an Econ would do.

Chapter 3 looks at the Player 1 anomaly and discusses why Player 1 does not behave like an Econ. We will discover that Player 1 is highly concerned with the watching eyes of others and reputational effects, and that he is quite sensitive to whether he will be perceived as good and just.

Chapter 4 looks at the Player 2 anomaly and discusses why she does not behave like an Econ. We will see how she is angered by unfairness and has a strong desire to punish opponents who act unfairly.

Then in Chap. 5, we return to Player 1's predicament and consider the faculties evolution imprinted on his brain and how these might explain why he was so compelled to worry about watching eyes and reputation. We'll also look briefly at some intriguing results that emerge from the ultimatum game.

Finally, in Chap. 6, we will look at how these anomalies represent a survival strategy implemented by modern humans' stone age brains, which I will refer to as **adaptive rationality**.

Now at long last, we are ready to delve into the details. We will begin in the next chapter by exploring exactly what sort of game the ultimatum game is and why it presents a puzzle.

## Chapter 1 Supplement

### Rationality

The following is a detailed note on the meaning of rationality for interested readers. Those eager to move on can safely skip ahead.

In economics, the most fundamental definition of **rationality** is that rational individuals have a consistent set of preferences about everything (Mas-Colell et al. 1995). Such individuals have clear preferences about which they are sure of in all scenarios, such that they never contradict themselves. For instance, a rational individual would not say: "I like strawberries more than peaches. I like grapes more than strawberries. But, you know, I really do like peaches more than grapes!" When a set of preferences is consistent and admits no such contradictions, it is said to obey the **transitivity** assumption. And when a set of preferences covers the entire set of alternatives (that is, the individual has a set of preferences about everything), it is said to obey the **completeness** assumption. When both these conditions are satisfied, says prominent game theorist Ariel Rubinstein (1998), the individual has *clear preferences*, and these are actually the only two conditions that need to be satisfied for rationality.

To dispel a possible misconception, I should note here that such individuals need not pursue their own self-interest exclusively. That is, they need not be **selfish**. Hence, this definition of rationality in no way conflicts with the existence of rational altruists, who are pleased when greater benefits accrue to others.

Yet traditional economics, at least, has assumed as a matter of course that individuals act rationally to pursue their own self-interest. Such an individual, the typical image of an Econ, is something that has come in for frequent ridicule. The great economist F.Y. Edgeworth wrote in his 1881 essay that "The first principle of Economics is that every agent is actuated only by self-interest." (Edgeworth 1881, p. 16). It is interesting to note that Edgeworth defines two species of action under which this principle operates: *war* and *contract*. The first refers not only to violence but to any sort of unilateral action taken without consent, while the second refers to cooperative action.

Werner Güth, the first to conduct ultimatum game experiments, noted in an early overview paper that "One reason why ultimatum bargaining became a widely known experimental paradigm is that experimentally observed ultimatum bargaining behavior clearly contradicts the most obvious *rationality* requirements of game theory and also of economic theory." (Güth and Tietz 1990, emphasis added). The word "rationality" has long been implicitly associated with selfishness, as explained by Henrich et al. (2004): "In the past, the assumption that actors were rational was typically linked to what we call the selfishness axiom—the assumption that individuals seek to maximize their own material gains in these interactions and expect others to do the same." The view was that it would not be reasonable for individuals to deliberately choose an option that they know will result in a loss to themselves. And no doubt this is why people were so surprised that ultimatum game experiments produced outcomes greatly different from what would be predicted were the participants to have pursued their own self-interest alone. Perhaps the reason the term "rational altruist" does not quite feel right to some is that the word "rational" is generally taken to imply a sort of dispassionate, even cold, concern for oneself to the exclusion of others.

It is worth noting that researchers (Dorison et al. 2022; Dorison and Heller 2022) have reported that, in the real world, rational individuals are not necessarily lauded by other members of society.

The economist Frank (1998, p. 2) notes that "There are almost as many definitions of rationality as there are people who have written on the subject." Indeed, various types of rationality have been proposed, including procedural rationality, substantive rationality, goal rationality and so on Shiozawa (2006). Incidentally, Frank's own definition of rational behavior is self-interested behavior.

It is against this backdrop that debate in economics these days draws a distinction between rationality and selfishness. It is argued, for instance, that the ultimatum game produces behavior that is rational but inconsistent with selfishness. In this book, I aim to uphold this distinction in the way I word my arguments, although without being too persnickety about it.

I will close with a modest defense of the Econ. I said earlier that the Econ might not be much fun to be around. Yet the Econ is also a good person, and in some sense an amazing one. The Econ does not get jealous. The Econ does not deprecate himself in comparison with others. He does not abandon hope for the future nor sink into despair. He is the kind of person who, no matter how hopeless the situation may seem, does not dwell on the past but calmly assesses the present, focuses solely on the future, and pursues what needs to be done with unwavering determination, choosing the best course of action from the options available. When I write it out like this, though, it does seem like a fairly unattainable standard for us.

For those intrigued by questions such as "What is rationality?" "Why are humans so smart yet so foolish?" and "Why does it matter?" be sure to read *Rationality: What It Is, Why It Seems Scarce, Why It Matters* (Pinker 2022).

# Chapter 2
# In Search of Homo Economicus

*Economics is a study of men as they live and move and think in the ordinary business of life.*
*—Alfred Marshall (1920), "Principles of Economics"*

## 2.1 Sharing with a Stranger

Now let's ease into the real topic discussion—the **ultimatum game**. I gave a basic overview of this game at the start of the book, but let's recap its basic structure. In simple terms, the game involves thinking about how to split money between two participants. It really is that simple, nothing that should cause any confusion.

**The ultimatum game: basic structure**

We first take a pair of people and designate one as Player 1, who will decide how to split the money, and the other as Player 2, who will receive a share of the money. Each sits in a separate room and will never see the other, not now or in the future. Neither has any idea of the other's name, appearance, or gender.

(Step 1) Player 1's proposal

First, we give Player 1 $10. We ask Player 1 to come up with a proposal for splitting that $10 between himself and Player 2, and again, Player 1 has no idea who Player 2 is. For ease of notation, Player 1's proposal can be written, for example, as ($6, $4), with the first amount going to Player 1 and the second amount going to Player 2 (Figs. 2.1).



K. Kobayashi, *The Mystery of the Ultimatum Game*,
https://doi.org/10.1007/978-981-97-8979-5_2

**Fig. 2.1** Ultimatum game—player 1's proposal

(Step 2) Player 2's response

Next, Player 2 decides whether to accept or reject Player 1's proposal. If Player 2 says "Yes", the money is split per Player 1's proposal, and each player pockets the cash (congratulations!). If Player 2 says "No", the $10 is forfeited, and neither player receives anything. This outcome can be written ($0, $0). Both Player 1 and Player 2 are well aware that this is how the game works (Figs. 2.2 and 2.3).

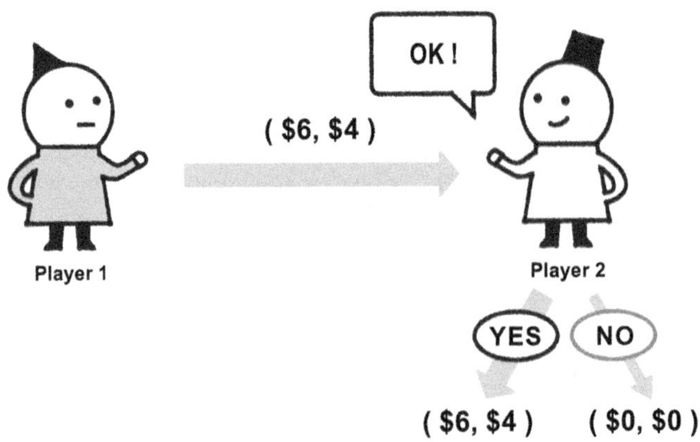


**Fig. 2.2** Ultimatum game—player 2 says "yes"

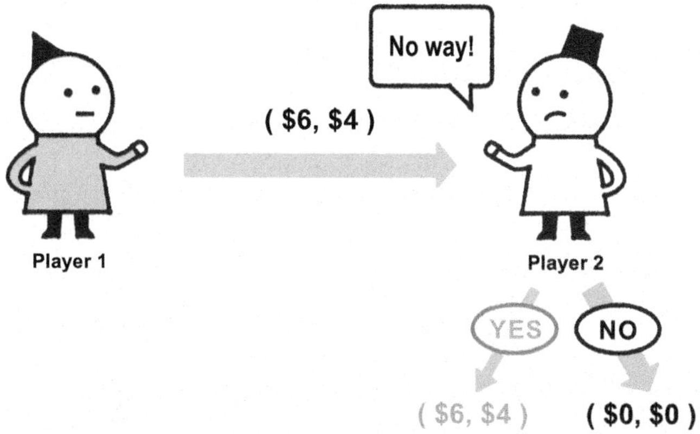


**Fig. 2.3** Ultimatum game—player 2 says "no"

That's all there is to the ultimatum game. Simple, right? Well, the fact that it's so utterly simple might leave some wondering what the point of it all is. In particular, the fact that nobody gains anything at all whenever Player 2 says "No" might seem somewhat strange or even wrongheaded.

This game was originally conceived as a simple model of bargaining behavior. That is, it was intended to be a bare-bones representation of bargaining situations that might occur in, say, a retail store. Something like this:

> Player 1: "Will you buy this for $xx?".
> Player 2: "OK" (successful bargain) or "No!" (no deal).

If Player 2 says "No", this signals that the bargaining process has broken down and no deal will be made—Player 2 will not buy the item—which is why this outcome was taken to mean that neither party makes any gain.

Subsequently, this game came to be widely adopted in a whole range of fields as a simple model for thinking about human conduct in society when altruism, fairness, and other related concepts are involved, such as when exploring questions like how much people are willing to share of what they already have with complete strangers, or what sort of split between oneself and another party people are willing to accept.

So, you should now understand the basic structure of the ultimatum game. Again, it really is quite simple, right? OK, so let's tackle some actual problems. What do you think will happen in the following scenarios?

**Econs under the scope: What would an Econ do?**

Let's start by looking at what happens when both players are Econs. This situation turns out to be easier to approach by considering how Player 2 would respond, so let's do that first.

‹Player 2's predicament›

(Case 1) Player 1 proposes ($5, $5)

Player 1 proposes ($5, $5). B's job here, as you now most certainly know, is to either accept or reject this proposal.

If Player 2 says "Yes", the payoff is $5. If Player 2 says "No", it's $0. An Econ would not hesitate to accept this proposal, because receiving $5 is unquestionably better than getting nothing at all (Fig. 2.4).

**Fig. 2.4** Ultimatum game: a $5 proposal

(Case 2) Player 1 proposes ($9.99, $0.01)

Let's say Player 1 is a little more ambitious (greedy?) and proposes ($9.99, $0.01). In this case, Player 2 stands to receive a grand total of one cent. What say you, Econ? Well, again, saying "Yes" and pocketing one cent is still better than saying "No" and walking away with nothing, so an Econ, in this situation, would not hesitate to accept the proposal (Fig. 2.5).

**Fig. 2.5** Ultimatum game: a $0.01 proposal

The astute reader (that's you!) will of course have no problem seeing where this leads. If you're an Econ, you will always say "Yes" whenever you stand to receive at least something (even the smallest possible amount of one cent) as opposed to nothing at all. One cent is still better than nothing to an Econ. And once this logic is in place, the rest is easy. Player 2, as the responder, will always agree to any proposal that provides a payoff of one cent or more.

‹Player 1's predicament›

Now you know exactly what Player 2 will do. So how about Player 1? If Player 1 is an Econ, what will he propose? I'm sure by this point it's obvious he will propose ($9.99, $0.01). Surely, reasons Player 1, Player 2 will agree to any proposal that leaves her with at least one cent. Player 1 has very little to ruminate over here. If a single cent is all that Player 1 needs to give up in order to get Player 2 to agree, his best course of action is clear: give Player 2 one cent and walk away with everything else!

Let's recap what we've figured out so far. If both Player 1 and 2 are highly analytical Econs, this is how the game will play out.

1. Player 2: Player 2 will always say "Yes" to any proposal that results in her pocketing at least one cent, since that is still better than saying "No" and getting nothing.
2. Player 1: Player 1 will anticipate Player 2's behavior and propose ($9.99, $0.01), thus expecting to take away the maximum amount the game will allow—virtually everything!

So if both players are Econs, Player 1 will, in quite rapacious fashion, say something like, "I'll take $9.99 and leave one cent for Player 2." And of course, Player 2 will agree to this because she still gets something rather than nothing. Ultimately, if both players are Econs, Player 1 should walk away with virtually everything[1] (Fig. 2.6).

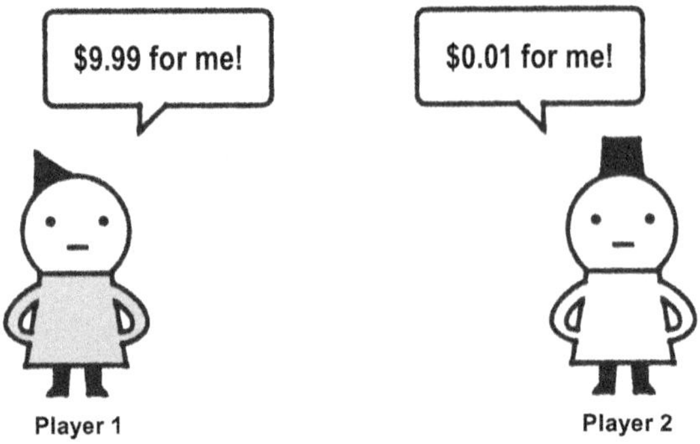


**Fig. 2.6** Ultimatum game: the end result

## 2.2 An Experiment: How Do People Really Behave?

**Everyday people under the scope—is this the most-conducted experiment ever?**

So far, we've learned that if both Player 1 and 2 are Econs, Player 1 should walk away with pretty much the entire stake. What do you think, though? Are you happy with Player 1 making off with the loot? It is at this point in my classes that the students will start to raise questions about the ultimatum game. "But if I were Player 2, there's no way I would say 'Yes' to a measly one cent! That's unfair!" they protest.

So how would an ordinary person, as opposed to an Econ, behave in this case? The only way to find out is to run the experiment!

Researchers have conducted this experiment in places like the US and Europe, as you might expect, and it has also been carried out in the interior of Africa, among peoples of the Amazon, and even in prisons. What's more, these experiments have not been exclusive to humans—researchers have even used the experiment to study the behavior of animals like chimpanzees. The experiment has been run in as many different locations and ways as you can think of, and so it is no wonder that, as previously mentioned, it is "claimed to be one of the most frequently performed of all human experiments today".

Let's see what the largest international comparative study has to say on the matter. Take a look at Fig. 2.7. This shows the locations of a huge series of experiments conducted by a team of eminent anthropologists and economists in 12 countries on four continents.

[1] Strictly speaking, there is one more answer to this problem. See the end-of-chapter supplement for details.

**Fig. 2.7** Location of experiments in small-scale societies. *Source* Henrich et al. (2004)

The range of societies in which the experiments were conducted is highly diverse. The researchers selected fifteen small, non-industrialized societies that continue to live by traditional means, encompassing foraging societies, societies that practice slash-and-burn horticulture, nomadic herding groups, and small-scale agricultural societies (Henrich et al. 2001, Henrich et al. 2004). This series of experiments was unprecedented in scale. Figure 2.8 gathers together the results of these experiments in small-scale societies, the results of experiments conducted in the US (Camerer 2003), and the results of pioneering international comparative research on Japan (Roth et al. 1991).

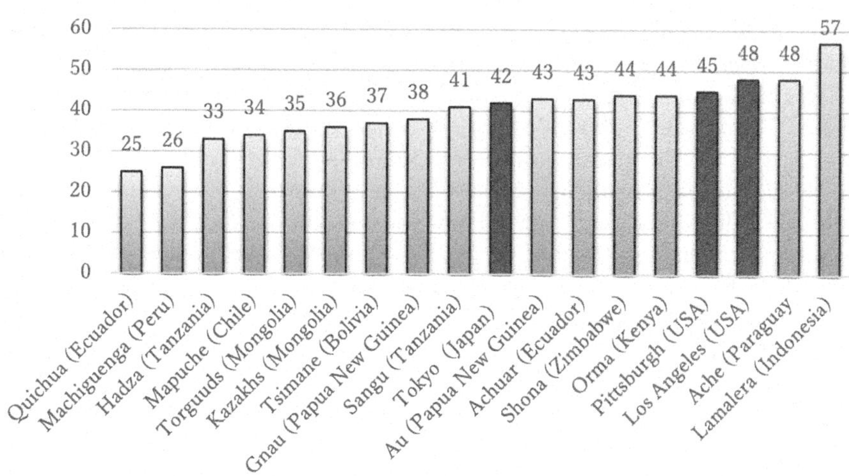


**Fig. 2.8** Average amount offered to player 2 as a percentage of the stake (lowest to highest). *Source* Prepared by the author based on Henrich et al. (2004) and Camerer (2003)

**Experimental results**

Figure 2.8 graphs the average amount that Player 1 offered to Player 2 as a percentage of the total stake in each location. Figures for modern industrialized societies like Japan and the US are shaded dark gray, while those for small-scale societies are shaded light gray.

The figure for Pittsburgh, USA, for example, appears as a dark gray bar just to the right of center. In this experiment, Player 1 offered Player 2 45% of the pot on average. This means that if the researcher provided a $10 pot, Player 1 offered Player 2 $4.50 on average.

Again let's recap quickly. Do you remember what percentage an Econ would propose to the responder? That's right, being an Econ, Player 1 would offer Player 2 the minimum amount of one cent. This is 0.1% if we're dealing with a pot of $10. Put simply, the proposer would attempt to walk away with virtually the entire stake.

Now look again at Fig. 2.8. The search was far and wide, yet nowhere did an experiment produce a figure even approaching 0.1%. In the vast majority of cases, Player 1 offered Player 2 40–50% of the pot, and fewer than five out of every 100 subjects offered less than 20% on average (Sigmund et al. 2002). The researchers did not find any societies (on the face of the Earth, at least) in which the standard behavior mirrors what economic theory asserts an Econ will do, which is to keep as much as possible for yourself.

**Culturally driven differences**

It's time to delve a little deeper into what the experiments revealed. The most common proposal in modern industrialized societies, like those found in the US and Europe, was to split the stake down the middle. The average proposals also offered around 40–50% to Player 2, which is close enough for us to say that the proceeds were generally split fairly evenly. Exceedingly low proposals were also rare, and such proposals were quite frequently rejected, which accords with the intuitions of my students as described earlier.

By comparison, the results were more varied in small-scale societies, as the graph shows (Henrich et al. 2004; Bowles 2004). The greatest disparity in the fraction proposed occurred among the Lamalera people of Indonesia, rightmost on the graph. Lamalera proposers, as the graph shows, tried to give their counterpart more than half. The authors of the study refer to such proposals as *hyper-fair*. The Lamalera earn their traditional living by hunting whales. They do this as a team, facing down whales with a single, hand-wielded harpoon in an extremely dangerous style of hunting. The hunters must work as a team, risking their lives to capture their prey, and the carcass is then divided according to strict cultural norms. The researchers point to this culture as a factor behind the hyper-fair behavior observed in this group.

Next to the Lamalera on the graph are the Aché of Paraguay. As one of the most generous groups, these people routinely offered their counterparts over 40% of the pot. The Aché culture also involves dividing up food resources obtained through foraging and hunting, such as meat and honey, equally among the group's members.

The Au and Gnau peoples of Papua New Guinea also often made hyper-fair offers. Yet interestingly, these hyper-fair offers, in which the proposer sought to give up more than half of the pie, were rejected roughly as often as proposals on the meager end of the scale. In these societies, people gain status through the unsolicited giving of gifts to each other, and accepting such gifts implies a strong obligation to reciprocate at some point in the future. That is, gift receivers owe gift givers a debt in some sense, and the researchers ascribe responder behavior to this cultural idiosyncrasy.

The closest any people came to behaving like Econs was among the Quichua and Machiguenga, which appear leftmost on the graph. The average proposal was quite small at around 25% of the pot, and moreover, most of the proposals were below 30%. What's more, those lowball proposals were almost never rejected. Among the Machiguenga in particular, the most common proposal was an exceptionally low 15%, with almost eight in ten proposals offering less than 30% of the pot. But despite the small amounts on offer, only one proposal was rejected, while the other proposals were accepted immediately (Henrich 2020).[2]

In four of the societies studied—the Tsimane, Aché, Kazakhs, and the aforementioned Quichua—Player 2 never rejected a proposal from Player 1. This may not be surprising in the case of the Aché, where almost all proposers offered 40% or more, but low offers were quite common among the Quichua and Tsimane.

So in some locations, subjects behaved more like Econs than people in modern industrial societies. It is also evident that people's behavior in the ultimatum game is influenced by cultural differences such as the degree of cooperation among individuals. Yet in all of the varied societies in which these experiments were carried out, the average fraction of the pot proposed by Player 1 was far removed from 0.1%, so the researchers were ultimately unable to find *Homo economicus* out there.

You may be interested to know that this vast, globe-spanning comparative work was triggered by the Machiguenga. The Machiguenga are a small-scale society of slash-and-burn horticulturalists living in the Amazon and have very little interaction with other societies.

As a young graduate student, the now renowned anthropologist Joseph Henrich ran ultimatum game experiments during a fieldwork trip among the Machiguenga and realized that they behaved quite differently from people in modern industrial societies. This somewhat serendipitous discovery led to a massive research undertaking. Henrich's paper discussing this work was titled "In Search of Homo Economicus". Those interested in Henrich's subsequent work are encouraged to read *The Secret of Our Success* (Henrich 2016) and *The WEIRDest People in the World* (Henrich 2020).

---

[2] However, the person who refused was not a member of the community; it was a relative visiting from another area (Henrich 2020).

**Experimental results**

Ultimatum game experiments have been conducted over and over again, in innumerable locations around the world beyond those so far mentioned here. Again, it is "claimed to be one of the most frequently performed of all human experiments today". And even after conducting ultimatum game experiments on a grand scale spanning the globe, we have been unable to find a group of Econs anywhere.

Accounts summarizing the results of these myriad experiments on the ultimatum game report the following general outcomes (Camerer 2003; Oosterbeek 2004).

Ultimatum game: experimental results (Figs. 2.9 and 2.10).

Player 1's behavior

- Most frequently offered fraction of the pot (mode): 40–50%
- Average proposal (mean): 30–40%

Player 2's behavior

- Often balks at low proposals (roughly 20–30% or less)

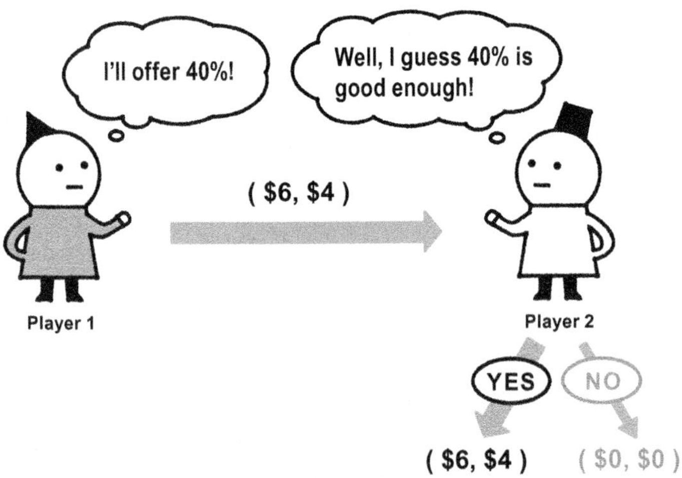


**Fig. 2.9** Ultimatum game: common outcome #1

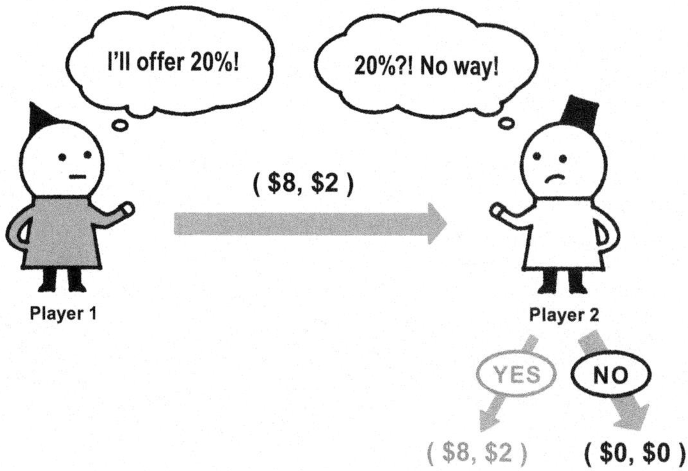


**Fig. 2.10** Ultimatum game: common outcome #2

What's so interesting about this experiment?

If you sit and think for just a little bit, it's easy to see what the most beneficial choice for you is. If someone says, "Here's $1 for free!" you will of course benefit by simply pocketing the cash. And it is not at all hard to see that you have ultimately made a loss if you make a point of knocking back that free money just because you think it's horribly unfair for the other person to keep $9 when you only get $1.

But even if you understand this simple logic, it still just doesn't feel right to acquiesce and say to yourself, "Well, sure, if I reject the offer, I lose the money, so I should be thankful for the $1!" It just feels wrong, and that's what the results of this experiment show. You can recite the logic until your purse strings fray, but that urge to reject lowball offers because they just feel wrong still wells up inside. "No way," you say, but clearly such behavior is unbefitting an Econ.[3]

[3] Some may be wondering whether changing the amount of money endowed to Player 1 (which we take to be $10 here) changes the way the players behave. This *endowment* is also commonly called the *stake*, and for the sake of lexical variety, you will also see it referred to as the *pot* or *pie* and so forth in the literature. While researchers have been unable to derive clear conclusions, it does appear that player behavior does not change much so long as the stake is not made quite considerably large (Andrea et al. 2019, Karagözolu and Urhan 2017).

However, it has also been reported that the average fraction proposed and the rate of rejection possibly both decline when the stake is larger (Cooper and Kagel 2016). To my knowledge, the biggest stake used in an experiment was slightly larger than the average annual income in the geographical region in which it was performed, so in US dollar terms, tens of thousands is probably roughly the sort of stake we are talking about (Andersen et al. 2011). In most experiments, the stake ranges from a few dollars to as much as a few dozen dollars, so the difference is one of orders of magnitude. In that experiment (the one with the highest stake), only one of 24 proposals was rejected. Just thinking about it casually, you can probably imagine someone rejecting 1% of $10, but you wouldn't expect too many people to reject 1% of $1 billion. At least, I know I wouldn't!

**Are those students in Cologne stupid?**

The first ultimatum game experiment was conducted by German economist Werner Güth among students at the University of Cologne (Güth et al. 1982). According to Güth, the fact that so many of the students sought to partition the money more or less fairly and utterly unlike what an Econ should do prompted some to ask "are those students in Cologne stupid?" (from an interview published on the website of the Max Planck Institute for Research on Collective Goods[4]).

In the same interview, Güth recalls how he received a lot of criticism for the way this experiment cast doubt over the notion of rationality. This should give you some idea of just how impactful the insight that humans do not always pursue their own narrowly focused self-interest was at the time.

## Summary

The ultimatum game

(1) How Econs behave (as predicted by economic theory).

- Player 1: Proposes to keep almost the entire pot.
- Player 2: Agrees to any proposal as long as she receives some amount, no matter how small.

(2) How real people behave (per experimental results)

- Player 1: Offers at least some appreciable amount to his counterpart (30–40% or so on average).
- Player 2: Rejects lowball proposals.

In sum, the results of experiments conducted around the world indicate that neither Player 1 nor Player 2 behaves like an Econ in practice. We are yet to discover *Homo economicus* anywhere.

How are we to wrestle with this problem? Let's break it into two component problems and address each separately.

Problem (1): The proposer's (Player 1's) predicament

"Why does Player 1 not try to keep (almost) the entire pot like an Econ would?"

Problem (2): The responder's (Player 2's) predicament

"Why does Player 2 reject some offers at the cost of losing something of value?"

The next two chapters will consider these two problems in turn. In Chap. 3, we'll focus on Player 1's predicament, and in Chap. 4, we'll look at Player 2's.

[4] In the same interview, Güth said that attempting to show that people do not behave the way that economic theory predicts would be like "overkilling an already dead man". Finally, it bears mentioning that the University of Cologne, established in 1388, is one of Europe's oldest universities and remains a highly regarded institution to this day.

# Chapter 2 Supplement

## The other solution

I explained somewhat earlier that the theoretical model solution to the ultimatum game is ($9.99, $0.01), which means Player 1 keeps $9.99 and Player 2 accepts one cent. That is, Player 1 proposes ($9.99, $0.01), and Player 2 accepts that proposal, and this is a **(Nash) equilibrium** (more formally, a **subgame perfect Nash equilibrium**).

There is actually one other (subgame perfect) equilibrium. This occurs when Player 1 proposes ($10, $0), meaning that he intends to take everything for himself, and Player 2 accepts this proposal. Let's run through why this is an equilibrium.

In game theory, the notion of an equilibrium has to do with thinking about whether you can benefit by changing your own actions based on the premise that the other player will not alter their course of action. In this case, Player 2 agrees to all proposals, even when she is allocated nothing at all. Now, you might be dubious about the prospect of someone agreeing to receive nothing. But think about it. If Player 1 proposes ($10, $0) and Player 2 says "No!" instead of "Yes", then Player 2 still does not benefit. That is, Player 2 cannot increase the benefit she receives by rejecting the offer instead of accepting it. So there is no point in Player 2 altering her choice as there is nothing additional to be gained from doing so.

Similarly, assuming that Player 2 will agree to all proposals, Player 1 cannot increase the benefit she stands to receive by changing the ($10, $0) proposal in any way. That is, Player 1 also has nothing to gain by changing his course of action. Hence, because neither of the players can increase their gain by choosing a different course of action, we have an equilibrium. More generally, in a game-theoretic equilibrium, none of the players will benefit themselves in any way by changing strategy.

Game theory reasons that when no one has anything to gain by altering their own course of action, then no one (or hardly anyone at all) will go out of their way to change their behavior, such that the status quo persists. Things remain as they are, and this is why such situations are called equilibria in game theory. Now, what is the inverse of this? Imagine a situation in which there is at least one individual who stands to benefit by changing his own behavior. Such an individual, it is reasoned, would be likely to act differently, and so the situation, as it stands, would not hold for long, and so such a situation would not fit the definition of an equilibrium.

The point for our discussion, ultimately, is that there are two equilibria. But since Player 1 walks away with virtually everything in both cases, there is not much difference between the two, and the choice of one over the other bears no influence on the fundamental insights to be gained from the discussion that ensues. As such, in this volume we'll couch our discussion in terms of the ($9.99, $0.01) equilibrium, since this is probably the more straightforward option.

That aside, there exists a book titled '*The Origins of Unfairness: Social Categories and Cultural Evolution,*' (O'Connor 2019), which discusses how issues such as gender differences and discrimination have evolved by adopting the concept of equilibrium. Those who are interested, please pick it up.

# Chapter 3
# Watching Eyes and Reputation Anxiety: Why Not Try to Take the Lot?

## 3.1 Ultimatum Game: An Offer You Can't Refuse!

**Fear of rejection?**

In this chapter, we consider Problem 1: Why does Player 1 not try to keep (almost) the entire pot like an Econ would?

When we looked at experiments conducted around the world, we did not find anyone who tried to keep almost everything to themself. Indeed, the vast majority of people in Player 1's position would probably be hard pressed to put forward such an ambitious proposal.

But why is that? Perhaps the first thing that comes to mind is the fear that Player 2 might very well reject your proposal if you try to hog everything. In fact, greedy proposals are fairly consistently rejected in experiments, and my students generally recoil at them too. "If that's all I'm offered, then no way!" they'll exclaim (Fig. 3.1).

**Fig. 3.1** Ultimatum game: fear of rejection



K. Kobayashi, *The Mystery of the Ultimatum Game*,
https://doi.org/10.1007/978-981-97-8979-5_3

If we hold this reasoning to be correct, it means that we believe Player 1 is factoring in the probability of Player 2 rejecting the proposal and making (what he believes to be) the minimum acceptable proposal on that basis (Fig. 3.2).

**Fig. 3.2** Ultimatum game: skirting the edge of rejection

Now, intuitively this way of thinking about the problem—the notion that Player 1 makes proposals on the basis of what he thinks Player 2 would find minimally acceptable—would seem to make sense. But doesn't it also involve a contradiction?

Put it this way: you're assuming that Player 1 will make a "reasonable" proposal while factoring in what Player B might be liable to refuse, but at the same time assuming that Player 2 might reject that proposal for illogical reasons such as an emotional "unfair!" response.

In other words, you're (arbitrarily) assuming that Player 1 behaves reasonably and Player 2 unreasonably. Is that really a reasonable approach to analysis of human behavior?

But let's set that problem aside for the moment and evaluate the idea that Player 1, fearing that Player 2 might say "no", makes the minimally acceptable proposal, just enough to escape rejection.

**For one with the lot, thanks!**

We'll proceed with this evaluation by considering a different game called the **dictator game**. If you're afraid of Player 2 rejecting your proposal, imagine being able to tie their hands, to stop them from saying "no". In the dictator game, the money is *always* divided according to Player 1's proposal.

In the ultimatum game, Player 2 gets to say either "yes" or "no". But in the dictator game, Player 2 no longer has any choice in the matter. Player 1 dictates how the money will be divided, and the money is always divided in accordance with Player 1's decision. Player 1 dictates what happens, and hence the name of the game (Fig. 3.3).

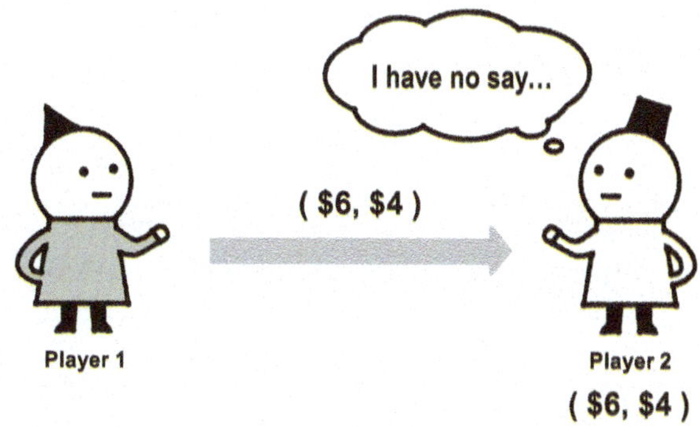


**Fig. 3.3** Dictator game

Pop quiz: If Player 1 is an Econ, what will he do?

That's right! You know it by this point! Player 1 simply takes the lot! If Player 1 was previously crafting his proposal under the shadow of Player 2's ability to reject it, and if the dictator game removes the ability for Player 2 to issue any such rejection, well then in the dictator game, Player 1 need not throw even a single cent Player 2's way. That is, Player 1 is free to propose ($10, $0) with impunity (Fig. 3.4).

**Fig. 3.4** Dictator game: the theoretical outcome

When you try this game out, however, that is not what happens. Try running it as a little thought experiment in your head. If you were in Player 1's shoes, would you suddenly leap up and say, "Ha ha! There's nothing they can do! I can just take the lot!"?

### Dictator game: Experiments

So how do real people behave in dictator game situations? There is only one way to find out. Once again, we turn to experiments!

Dictator game experiments have been conducted globally and extensively enough to be called on par with the use of ultimatum game experiments. Researchers (Camerer 2003; Engel 2011) have summarized the results of over 600 such experiments conducted around the globe as follows.

Dictator game

Player 1's behaviour

- Average fraction of the stake offered: Around 20–30%
- Fraction of people who give anything at all to Player 2: Around two in three.
- Fraction of people who split the stake evenly: Around one in six.

Thus, Player 2 receives some 20–30% of the stake in the dictator game. In the ultimatum game, Player 2 generally received 30–40%, and so although the amount received is lower here, it is not zero. One in six people even split the stake fifty-fifty, so evidently people's behavior in this game is completely different from what we would expect of an Econ. This has even been called "One of the most puzzling results in experimental game theory" (Franzen and Pointner 2012).

What we learn from this result is that people's behavior in the original ultimatum game has something of a strategic aspect to it. Player 1 wants to take as much for himself as possible, but if he tries to take too much and Player 2 rejects the proposal, he ends up losing everything. The key to increasing the benefit to yourself must lie in somehow figuring out just how much you can increase your share of the pot without the other player saying "no". It seems evident that people do in fact adopt this sort of highly strategic mode of behavior (Fig. 3.5).

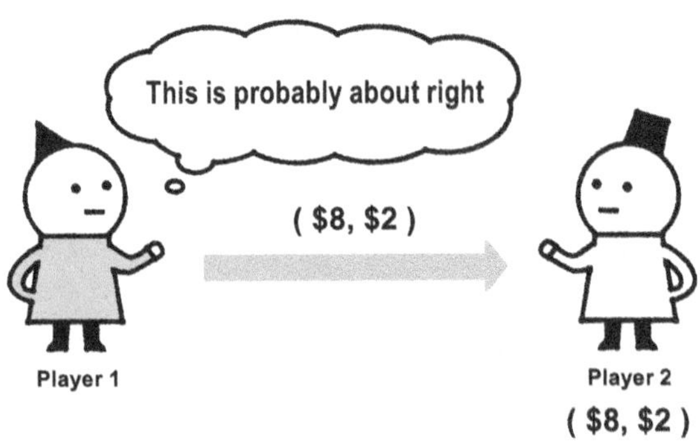


**Fig. 3.5** Dictator game: the experimental result

**The faint 20% sliver of hope**

Neurologist Naotaka Fujii casts this dictator game result in a hopeful light (using a phrase that translates literally to "the 20% hope"; Fujii 2009). If we are capable of giving as much as 20% of what we have to a complete stranger, he reasons, then perhaps we could do something amazing if we were to pool this money together as a society.

What we really want is to get at the innards of this 20%, to figure out why someone would give up 20% of what they have been given for someone they know nothing about.

Some people might say this is altruism. Indeed, the fractions allotted to others in the dictator game have been used in a range of studies as a measure of human altruism. In many of those experiments, the players were complete strangers with no idea of each other's identity. Perhaps it is possible to see the act of giving 20% of what you hold to a complete stranger as a beautiful act of altruism. But is this faint 20% sliver of hope really entirely a reflection of altruism?

Why does this apparently altruistic behavior of giving money to strangers occur? In other words, what lies at the heart of that potentially game-changing 20%? From here on out, this is the question we will be focusing on.

What exactly is altruism? This question is probably a bit daunting to tackle head-on, so we'll start by coming at it from a different angle.

## 3.2 Watching Eyes: I See You…

**Anonymity: If you can't get caught…**

Let's review the experimental evidence so far. In both the ultimatum and dictator games, the players were almost always complete strangers with respect to each other. But hang on. There was at least one person who knew precisely how much money was changing hands. Now, if it's not immediately obvious who that person is, pause and think it over for a moment.

Got it?

Yes, the researcher! When these experiments are conducted at a university, the student body is often tapped to play the game, and the researcher is often one of the students' lecturers. Imagine yourself as a student in this situation. While your identity may be completely shielded from the other player, you might still feel pretty uncomfortable about your lecturer looking over your shoulder. Even so, is that alone really enough to alter the outcome? There's only one way to find out. Time for an experiment!

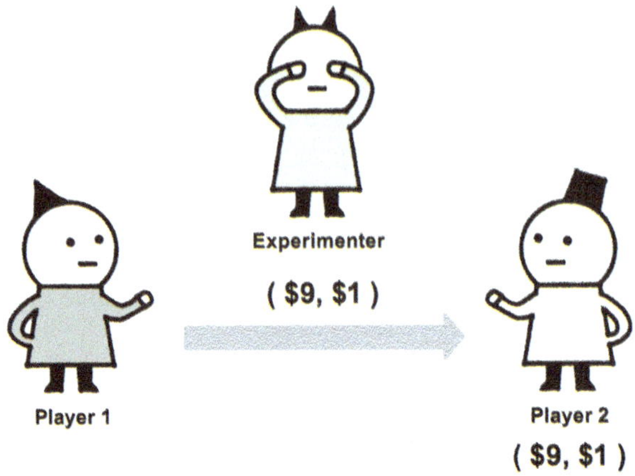


**Fig. 3.6** Dictator game: blinding the expeimenter

The dictator game experiment depicted here (Fig. 3.6) was conducted in such a manner as to ensure that both the other player and the researcher would be blind to Player 1's actions (Hoffman et al. 1994). The results were surprising. Some 60% of people in Player 1's position gave nothing to Player 2 and kept everything to themselves. Figure 3.7 shows the overall breakdown. On average, around 10% of the stake was offered to Player 2, a figure of less than half that observed with the ordinary dictator game setup. The act of the researcher simply observing the proceedings most certainly seems to have influenced how people behave.

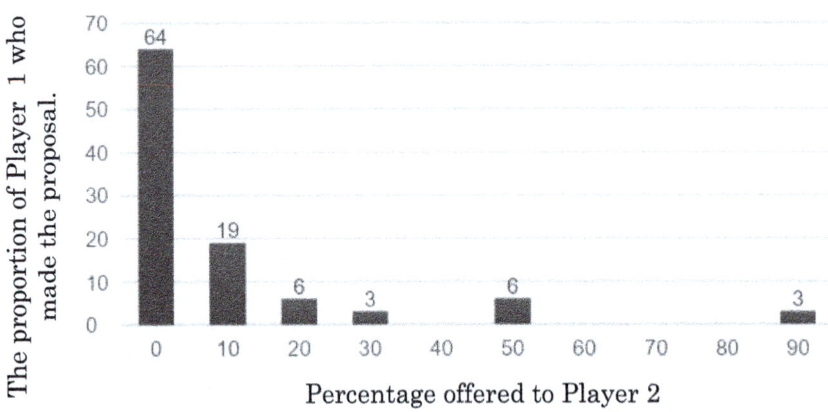


**Fig. 3.7** What percentages of player 1 offered what percentage shares to player 2 when the researcher (experimenter) was blinded? *Source* Hoffman et al. (1994), (For ease of comparison, monetary amounts have been converted to proportions)

But wait again. We still have one other problem to consider here, and it's a little esoteric. So the experiment was dutifully set up to ensure objective anonymity, such that neither the researcher nor anyone else would know what either of the players chose to do. And yet, from the perspective of a player in this game, could you really have complete confidence that no one, not even the researcher, would have any idea what you chose to do?

Diving deeper here, we encounter something called the *randomized response technique*, which researchers have used to buttress the assurance of anonymity in these games. Follow-up dictator game experiments have used this technique to make it patently obvious to the players that no one would have any idea, whatsoever, of their choices.

And quite surprisingly yet again, the average fraction apportioned to the other player in these experiments was lower even still, at a miserly 7% of the total stake (Franzen and Pointner 2012). In terms of our standard yardstick stake, this means Player 1, on average, only gave $0.69 to Player 2 out of an entire $10. A paltry 69 cents! Yet this is all quite reasonable, wouldn't you say? The less you feel you are being watched, and hence the lower you perceive to be the likelihood that your behavior will be noticed or exposed by anyone, the more you might be (subconsciously) drawn to change your behavior in the direction of your own interests.[1]

Other studies have also shown that human behavior is influenced by the presence of an observer. For instance, the quality of goods that dealers deliver is lower when no one is watching in a real-world market (Levitt & List 2008), and conversely, when observers are present, people more readily engage in volunteer activities (particularly volunteering that is costly—not necessarily in monetary terms—to the individual concerned), and people are more productive at work when they know they can be seen by highly productive peers (Mas & Moretti 2009). This sort of effect on behavior is termed the **audience effect**. Experiments have even shown that darkness (in the sense of dimmed lighting, for example) can prompt people to cheat more and act more selfishly (Zhong et al. 2010). We humans, for better or worse, must admit that this sort of thing is part of our psyche.

### Watching eyes: Lab experiments

Alright, so we know that people's actions can be influenced by whether they are aware of whether or not they are being watched. To put it simply, when someone is watching, people behave in a way that is less likely to attract criticism, and when no one is watching, people gravitate toward serving their own interests. Still, all of this simply means that you will probably behave properly if someone else is looking, which sounds like garden-variety common sense. Fair enough. But even more interesting

[1] It must be noted, however, that while others have reported similar findings about the effect of anonymity or the presence of the researcher's "watching eyes" (List et al. 2004; Thielmann et al. 2016), still others have found that the effect does not necessarily show up (Bolton et al. 1998; Barmettler et al. 2012).

results have also been reported, results that perhaps suggest the "watching eye" need not even be real.

Researchers performed a dictator game experiment in which the players made their decisions in the presence of a computer displaying stylized eyespots—images designed to evoke the sense of eyes watching (Haley and Fessler 2005). The game was played by 248 undergraduates aged 19–36 years at the University of California, Los Angeles. All the researchers did was place a computer displaying these stylized eyespots in front of the players. That's all. Yet, perhaps surprisingly, this had a substantial impact on the results.

Take a look at Fig. 3.8, which shows the amount of money allocated to Player 2 in this dictator game experiment. The amount allocated increased a startling 30% when the stylized eyespots were on display. Figure 3.9 shows the actual computer displays used, eyespots on the left and the control display on the right. You may recognize the eyespots as the eye of Horus, a symbol of the ancient Egyptian god.

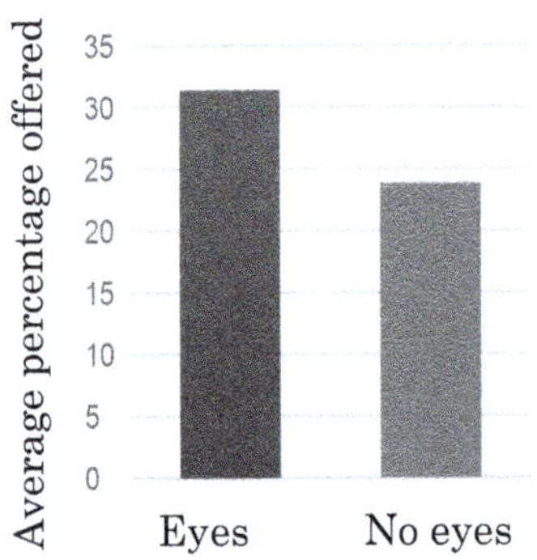


**Fig. 3.8** *Source* Haley and Fessler (2005)

**Fig. 3.9** The computer displays used in the experiment. *Source* Haley and Fessler (2005)

Average fraction allocated to Player 2 when the watching eyes were and were not present.

These findings—that people react to the presence of fake eyes—may seem hard to believe at first, but they have subsequently been corroborated by a bevy of work from other researchers, who have showed, for instance, that in the presence of watching eyes, people become more generous (Rigdon et al. 2009; Nettle et al. 2013; Mifune et al. 2010; Oda et al. 2011; Fenzl and Brudermann 2021), more socially cooperative (Burnham and Hare 2006), lie less (Oda et al. 2015), and condemn moral violations more severely (Bourrat et al. 2011; Li et al. 2021). Others have also shown that this sort of effect holds even in the presence of robot eyes, so it is not limited to the presence of stylized images or photographs of human faces (Burnham and Hare 2006).

More fascinating still is the finding that this sort of effect can be observed when mere black ink dots are appropriately arranged (Rigdon et al. 2009). Have a quick look at Fig. 3.10. Three dots arranged like a pyramid (.·.) did not produce an effect, but when flipped up to resemble a face (·.·), the researchers observed the same sort of effect as described above in male participants. Even a rudimentary representation of watching eyes like this may cause people to alter their behavior without even realizing it.[2]

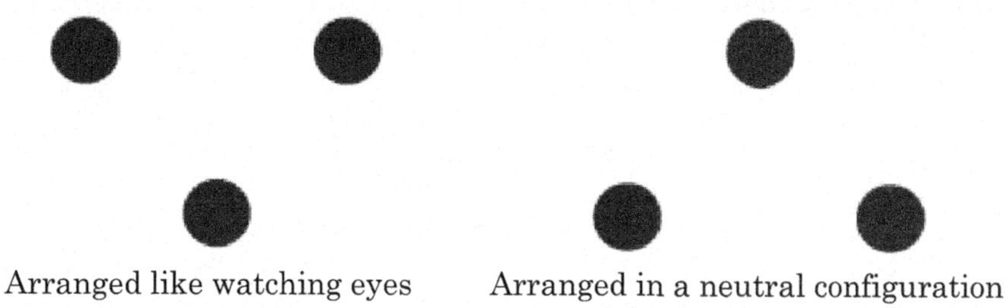


**Fig. 3.10** Three dots arranged in triangle formations. *Source* Rigdon et al. (2009)

**Lab and field experiments**

Here a problem arises. If the presence of watching eyes can alter people's behavior, then the way we perform experiments is, itself, problematic.

When a researcher gathers subjects in a classroom or other venue and says, "Let's do an experiment," the experiment that ensues is called a **laboratory (lab) experiment**. The experiment that used eyespots displayed on a computer screen discussed above is an example of a lab experiment.

[2] It has also been reported that these three-dot eyespots have an effect in field experiments (Krupka and Croson 2016), which we will discuss shortly. Researchers have also shown that newborn babies respond to three-dot eyespots (·.·), especially in the case of black dots on a white background (Farroni et al. 2005), and that unborn fetuses may even respond to face-like visual stimuli delivered in the form of red light shone through the mother's abdomen (Reid et al. 2017).

In these sorts of experiments, it is clear to the participants that there is a researcher who is observing their behavior under the pretext of an experiment, no matter how anonymous the researcher attempts to make things. Unless you could somehow prevent the participants from realizing that they are part of an experiment, then concealing the existence of the researcher's watching eyes from the participants seems like a tall order. Is it even possible to perform an experiment in such a way that the participants are unaware that the experiment is taking place?

**Watching eyes: Field experiments**

Well, consider the following experiment (Bateson et al. 2006). To set the scene, an honesty box had long been in place in a certain university coffee room to collect payment for drinks. The researchers posted an image banner on a cupboard door above the honesty box and changed the image once a week. They then compared the rates of money collected according to image type. This is a famous experiment and already regarded as a classic in the field.

The "participants" came from a population of 48 members of the Division of Psychology at the University of Newcastle. They simply came to the coffee room as normal, took their drinks, and deposited the money for those drinks in the honesty box. The image banner was changed weekly, but the participants had no idea that the researchers were monitoring how much money was received in the box for each image type.

With this method, participants are unaware of having been roped into an experiment, and thus there is no reason for them to worry about a researcher observing their behavior. Experiments performed in more natural or real-world settings like this are called **field experiments**. They are performed "in the field", which simply means not in an artificially constructed environment like a lab.

Anyway, all the researchers did was change an image posted on the coffee room cupboard, so you might be puzzling over how this could possibly have any influence on people's behavior. Surprisingly, however, the researchers found that when an image of human eyes was posted instead of an image of flowers, payment rates jumped to around threefold higher. You can clearly see from Fig. 3.11 that the amount paid zigzags nicely back and forth according to banner image type.

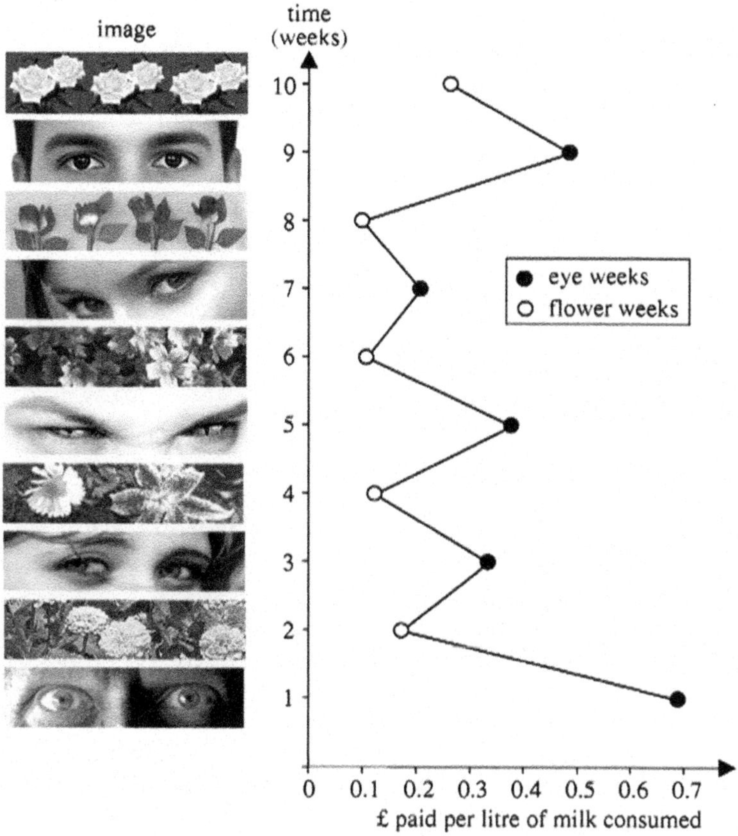


**Fig. 3.11** Amount paid by image type. *Source* Bateson et al. (2006)[3]

## Watching eyes: Subsequent field experiments

The finding that human behavior can be influenced to such an extent by something so seemingly inconsequential as an image of eyes posted on a cupboard door certainly created a stir. A slew of experiments in this vein ensued, with researchers reporting a smorgasbord of varied results. When "watching eyes" are posted in some location, for instance, people refrain from leaving litter on cafeteria tables (Ernest-Jones et al. 2011), people litter less in general (Bateson et al. 2015), people spend more time removing garbage (Francey and Bergmüller 2012), people donate more (Powell et al. 2012; Krupka and Croson 2016; Oda and Ichihashi 2016; Kelsey et al. 2018), bicycle

[3] Tea, coffee, and milk were all available in the coffee room, but the researchers used the volume of milk consumed each week as an index of total beverage consumption.

theft decreases (Nettle et al. 2012),[4] voter turnout increases (Panagopoulos 2014), and people wash their hands properly (Pfattheicher et al. 2018).

We've had evidence for some time now that people are quick to realize when someone is looking straight at them (Von Grünau and Anston 1995), and this finding holds up when the subjects are children as well (Senju et al. 2005).[5] Researchers have also found that when we perceive someone to be paying visual attention to (looking at) us, this causes us to pay attention to them (Colombatto and Scholl 2022). Others have reported that we more readily remember faces that are looking directly at us (Hood et al. 2003; Mason et al. 2004). Watching eyes can also induce a sense of being seen (Pfattheicher and Keller 2015) and evoke negative emotions such as anxiety, distress, and nervousness (Panagopoulos and van der Linden 2017). Meta-analyses combining the results from multiple studies have also confirmed the watching-eyes effect (Dear et al. 2019, Wang et al. 2023).[6]

When someone looks at you, it means they are taking some sort of interest in you, and this is an important social signal. The human brain has the ability to quickly detect a direct gaze and to accommodate social information and adjust social behavior accordingly (Senju 2012). The idea that eye contact influences social cognition and behavior in various ways is also called the "eye contact effect" (Senju & Johnson 2009).[7]

We have long known that mirrors can have effects similar to those induced by eyes. The presence of a mirror in which people can see themselves can reduce cheating or unfair behavior, for instance (Beaman et al. 1979, Batson et al. 1999). Some studies have also shown that people may be more prosocial, less deceptive, and more generous in response to religious priming designed to make them conscious of God—having subjects recall religious symbols or read sentences containing words

---

[4] The researchers studied bicycle theft on a university campus. Although bicycle thefts decreased at locations where images of "watching eyes" were posted, thefts elsewhere on campus increased, resulting in little overall change in the number of bicycle thefts across the campus as a whole. Hence, it seems likely that the images simply displaced offending to different locations.

[5] Note also that the literature discusses the possibility that people have a special sensitivity to eyes and faces (or face-like objects) from a very young age, with newborns paying more attention to face-like shapes than to other objects. Researchers have found, for instance, that human infants only a few days old prefer to look at faces that engage them in a mutual gaze rather than averted faces (Farroni et al. 2002) and that four-month-old infants exhibit strong brain activity (cortical activation) when presented with faces looking directly at them as opposed to an averted gaze (Grossmann et al. 2008).

[6] It should be noted, however, that while people may respond involuntarily to watching eyes, the effect may be much more limited than originally thought (Kobayashi mimeo). And the effect reported in Wang et al. (2023) was by no means large. Some meta-analyses have even found there to be no watching-eyes effect (Northover et al. 2017a, Northover et al. 2017b). See the end-of-chapter supplement for details.

[7] It has been pointed out, however, that the way in which the eye contact effect manifests itself may be influenced by the way facial stimuli are presented and the task at hand (Senju 2012). This is related to footnote 10 above. Brain research is showing that humans do not necessarily react in the same way to eye contact in all contexts and situations.

related to God[8] (Shariff and Norenzayan 2007; Shariff et al. 2016). This approach perhaps evokes a sort of "inner eye" in the subjects.[9]

Subsequent research into the details has also shown that the watching-eyes effect not only increases the average amount of money allocated to the other player but also influences the decision to give any money at all. This finding is not concerned with whether people who give money in the first place give more in the presence of watching-eyes but, rather, it says that people who otherwise want to hog the stake will actually give the other player something when eyes are watching (Nettle et al. 2013). In other words, rather than making people more generous in general, the watching-eyes effect may be prompting people to avoid reputational damage by doing the bare minimum to stay within (not transgress) social norms.[10]

**Eyes of a predator**

From the discussion so far, we know that people may react in various ways to the presence of watching eyes. This special sensitivity to eyes and the direction of someone's gaze may not be unique to humans, however (Endo 2005).

Primates like chimpanzees and baboons are not the only animals to exhibit eyespot-avoidance behavior. It has also been observed in birds like crows (Kobayashi 2019). Research has also shown that predatory birds are less prone to attacking butterflies with eyespots across their wings, so eyespots can increase prey survival (Kodandaramaiah et al. 2009). And quite intriguingly, researchers have shown that eyespots painted on cattle rumps are associated with reduced attacks by ambush predators such as lions (Radford et al. 2020). This result comes from a four-year field experiment involving over 2,000 head of cattle.

In the words of evolutionary biologist Maestripieri, "For some animals, detecting the eyes of a predator means the difference between life and death" (Maestripieri 2012). This is because eyes looking directly at you can mean you are in danger of being attacked by a predator. This special sensitivity to eyes staring directly at us

---

[8] Note that these results are not robust among non-religious participants (Shariff et al. 2016; Lin et al. 2016). Note also the possibility of cultural differences (Miyatake and Higuchi 2017) and concerns about reproducibility (van Elk et al. 2015; Purzycki et al. 2016).

Research also shows that people who believe in gods that have knowledge about human thoughts and actions, and that sanction violators (punitive gods), may behave in a fairer manner toward others (Purzycki et al. 2016).

[9] This inner eye may also be related to the "culture of sin" predominant in the West and "culture of shame" predominant in the East (Benedict 1946). We know that people in different cultures not only behave differently but also think about the world differently, both cognitively and emotionally, and even exhibit differences in hormone secretion and brain responses (Nisbett and Cohen 1996; Nisbett 2004; Henrich 2017). I have provided very little discussion of cultural factors owing to space limitations, but I think it's clear culture has played an extremely important role in evolution (see, e.g., Henrich 2017).

[10] Similarly, Kawamura & Kusumi (2017) and Fenzl & Brudermann (2021) argue that the effect of watching eyes is to prompt people to avoid bad reputation. It has been argued, meanwhile, that cues of being watched, rather than evoking a fear of punishment, may prompt people to anticipate a reward in the form of good reputation (Oda et al. 2011). Thus the debate continues as to whether people seem to be concerned about watching eyes because they want to avoid a bad reputation or because they want to pursue a better reputation.

may have originated as a way of quickly detecting danger in response to potential predation.

## 3.3 Bonds and Loneliness, a Carrot-And-Stick Affair

**A good reputation is scrumptious!**

We have so far looked at the possibility that humans and other animals are highly sensitive to the eyes of other beings. For us humans, this sensitivity to the watching eyes of others is perhaps evidence of just how much we care about what others think of us, about our reputations. So what significance do these things have in the human brain?

When we eat tasty food or receive money, our brain's **reward system**, the system that responds to joy and pleasure, lights up (especially the **striatum**, a particularly important part of the reward system. The striatum corresponds to the area shaded in the Fig. 3.12. These brain responses can be interpreted simply as "getting money makes me happy!" or "eating good food makes me happy!" We can also say that it is precisely because it makes us happy that we feel a desire to get more money or eat more tasty food.

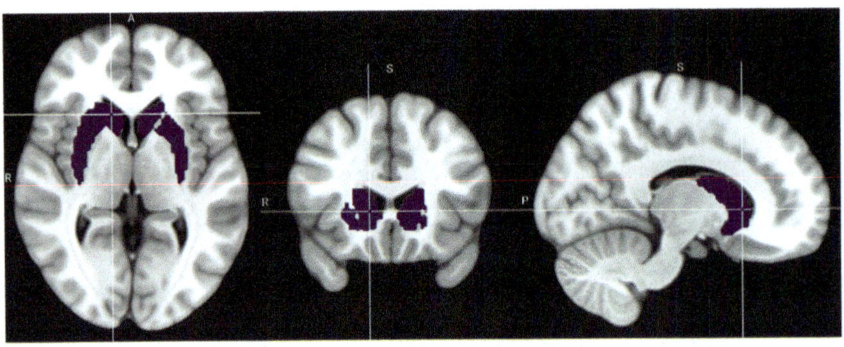

**Fig. 3.12** Striatum (image provided by professor Asamizuya, created using MRIcroGL)

Others' opinions of us and our reputations, meanwhile, are intangible. We cannot eat them, and we do not immediately gain or lose anything tangible as a result of them. Yet experimental evidence shows that our reward system (striatum) responds to intangible "social rewards", like compliments and praise, in the same way as it does to physical rewards like money (Izuma et al. 2008). Moreover, we know that receiving more money produces a stronger reward system response, and this response holds up for intangible rewards as well: the higher the praise received, the stronger the reward system response. This makes sense when the praise comes from the people who matter most to us, but research also indicates that the brain's reward system is activated in response to positive feedback from complete strangers as well

(Lieberman 2013b). In short, we know that praise and positive feedback are just as *real* to the brain as good food and money (Fig. 3.13).

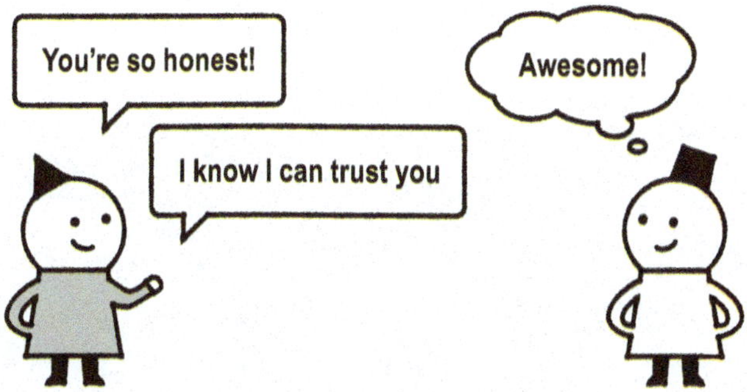


**Fig. 3.13** Others' opinions of us

### Rejection "hurts"

We also know that humans are very sensitive to exclusion from groups.

Imagine you have control of a character on a computer screen and your character is engaged in a virtual ball-tossing game with two other participants (Eisenberger et al. 2003). This is known as the Cyberball experiment. Unbeknownst to you, however, the other two participants are part of the study. They are collaborators. Now imagine that these two collaborators toss the ball between themselves. The ball never comes to you. You're suddenly and summarily excluded from the group (Fig. 3.14).

**Fig. 3.14** Cyberball

When the researchers conducting this experiment measured excluded subjects' brain responses, they found that areas associated with actual physical pain—the **anterior cingulate cortex** (ACC) (Fig. 3.15a) and the **anterior insula** (Fig. 3.15b)—were activated. Moreover, individuals who reported stronger feelings of rejection showed greater activity in the ACC.

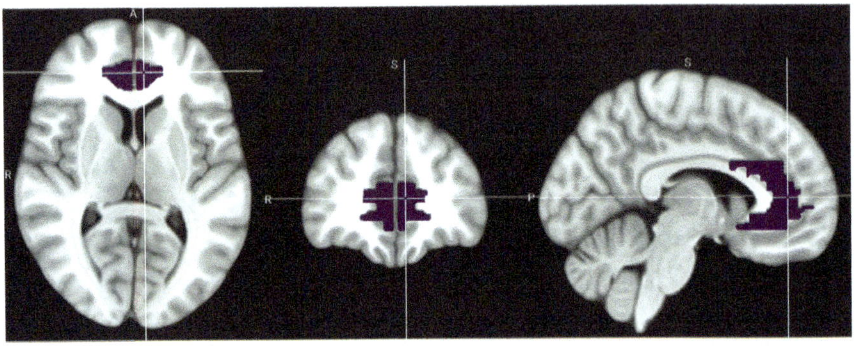

**Fig. 3.15a** Anterior cingulate cortex: ACC (Image provided by professor Asamizuya, created using MRIcroGL)

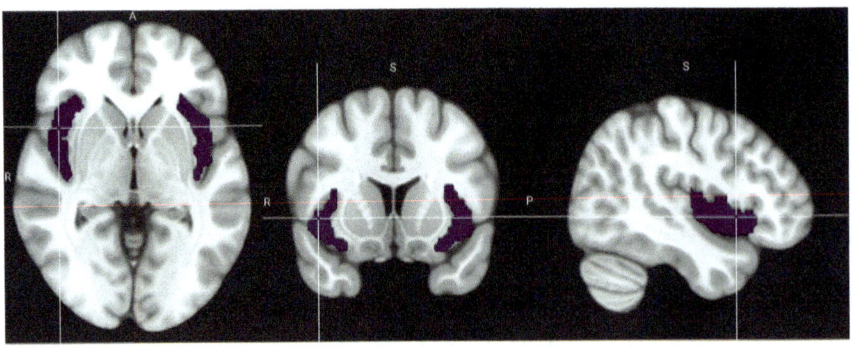

**Fig. 3.15b** Insula (Image provided by professor Asamizuya, created using MRIcroGL)

This study was carried out by the famous social neurologist Matthew D. Lieberman and colleagues. As Lieberman recalls it, he was analyzing the data with his colleague one night in the lab, and by coincidence one of his graduate students was analyzing data from a physical pain study on the next computer. When they compared the two computer screens, they realized that the brain's responses to physical pain and to social pain were strikingly similar (Lieberman 2013b).

The participants in this study did not know each other at all, and they would not be meeting up with each other afterwards, so you might think, "Who cares?" Not being tossed a virtual ball in a virtual environment does not result in any sort of tangible loss. And yet, to the brain, this sort of exclusion feels like pain. "Heartache" is not merely a figure of speech, it seems.

Other research suggests that the brain exhibits a similar pain response to this sort of ostracism in a range of scenarios. This sort of effect shows up even when people are ostracized by out-group members, that is, people who are not members of their own in-group (Williams et al. 2000), and even when the ostracism comes from members of a despised group (Gonsalkorale and Williams 2007). It even shows up when the participants know full well that they are playing against a computer and not a real person (Zadro et al. 2004). In short, humans are highly sensitive to exclusion from groups.[11]

The social psychologist Kipling D. Williams was the first to perform a Cyberball experiment (Williams et al. 2000). One afternoon while relaxing in a park, Williams noticed a Frisbee roll up beside him. He threw it back to the two strangers who had evidently made the errant throw, prompting a spontaneous game of three-way toss. Minutes later, the strangers continued their game without him. It became apparent to Williams that he would no longer be getting the Frisbee. While this may seem like perfectly natural behavior, Williams recalls that he felt a strange sense of alienation, and that's when he came up with the idea of Cyberball to study ostracism in the lab.

Many Cyberball experiments have since been conducted, with most of them involving an experience lasting only a few minutes with strangers in a virtual space. Even so, the effects were surprisingly strong, with ostracized subjects enduring feelings of sadness, anger, and alienation and experiencing negative effects on their sense of control, self-esteem, and even their sense of meaningful existence (Williams and Nida 2011).

Williams, the originator of Cyberball, muses, "How can it be that such a brief experience, even when ignored and excluded by strangers with whom the individual will never have any face-to-face interaction, has such a powerful effect?" (Williams and Nida 2011). Why are people so negatively affected by exclusion and ostracism?

**Loneliness in an evolutionary context**

The ability to feel physical pain helps us humans avoid physical danger. Renowned psychologist John T. Cacioppo, who studied loneliness, argued that the feeling of loneliness evolved for similar reasons (Cacioppo and Patrick 2008). Simply put, "loneliness developed as a stimulus to get humans to pay more attention to their social connections, and to reach out toward others, to renew frayed or broken bonds." In other words, the ability to feel the pain of loneliness helps humans avoid the dangers inherent in being isolated.

[11] However, research has shown that people who receive daily social support exhibit a weaker pain response to temporary social exclusion, such as in the Cyberball experiment (Eisenberger et al. 2007). On the flip side, this may mean that people who do not receive much social support on a regular basis and thus constantly feel lonely experience a stronger sense of pain even to very brief social exclusion.

Subsequent research has also shown that those who comply with instructions to ostracize strangers for no apparent reason felt "pain" in the same way as when being ostracized (Legate et al. 2013).

In the words of developmental psychologist John Bowlby, known for his pioneering work in attachment theory, "To be isolated from your band… and especially when young, to be isolated from your particular caretaker, is fraught with the greatest danger. Can we wonder then that each animal is equipped with an instinctive disposition to avoid isolation and to maintain proximity?".

Lieberman (of previously discussed Cyberball fame) argues that in mammals like us, for whom isolation from the herd or tribe can be perilous, social pain appears to have hijacked the brain's physical pain system as a means of sounding the alarm when our social bonds are threatened, since this in turn represents a threat to our survival (Liberman 2013b). Pain is the body's way of crying out, "Avoid whatever is causing this!".[12]

**Interpersonal bonds are the carrot, loneliness is the stick**

Let's recap what we've learned. The social joy we derive from praise and positive feedback and the social distress we feel when others treat us as a nuisance or exclude us produce responses in the brain that are highly similar to those produced by real and tangible things like money and material possessions.

Research on the brain has illuminated the possibility that a mere desire to be seen as a good person or to avoid being left out of a group, quite separate from any potential monetary or material reward, is enough to motivate great expenditures of effort in people. As Cacioppo says, "the warmth of connection is the carrot; the pain of feeling isolated, also known as loneliness, is the stick" (Cacioppo and Patrick 2008).

**A brief aside on loneliness**

The problem of loneliness has received much attention in recent years. In 2018, the United Kingdom became the first to appoint a ministerial lead for loneliness (the position is often referred to colloquially as Minister for Loneliness), having concluded that loneliness is one of the biggest public health crises of our time.

Prolonged solitude cut off from social connections is extremely stressful for us humans (Cacioppo and Patrick 2008; Murthy 2020; Dunbar 2021). It impairs immune and cognitive function (Cacioppo and Patrick 2008) and is linked to an alarming range of health risks, including dementia (Freak-Poli et al. 2022), depression, heart disease, stroke, and premature death, and researchers have found that its effects can be severe enough to ultimately threaten lives (Holt-Lunstad et al. 2015; Valtorta et al. 2016; Rico-Uribe et al. 2018; Russo 2018; Bu et al. 2020). The negative impact of loneliness on health is now said to be even greater than that of smoking, obesity, excessive alcohol consumption, and lack of exercise (Murthy 2020).

---

[12] A fairly recent study found, perhaps not surprisingly, that people seem to crave social interaction after isolation (Tomova et al. 2020). After a period of mandated social isolation, people were shown pictures of social contact cues, such as people joking around with friends. The reaction this evoked in the brain was quite similar to the reaction evoked by pictures of food viewed after a period of fasting. Another previous study showed that people feel "pain" when observing other people being ostracized that is similar to what they feel when they themselves are ostracized (Giesen and Echterhoff 2018).

Why does this happen? It is believed to have to do with changes in the body to protect itself (in the short term) from enemies. Upon detecting signs of loneliness, our entire body goes into alert mode, rapidly increasing stress hormones, dilating pupils, and increasing the heart rate so that the muscles move smoothly for fighting or fleeing. We become sensitive to signals around us so that we can quickly detect the presence of enemies, paying no attention to extraneous thoughts. Furthermore, our sleep becomes shallow and fragmented so that we will not be attacked by enemies while asleep. Yet, should this state persist for a long time, the heightened levels of stress hormones exert considerable pressure on the cardiovascular system and can lead to damage to tissues and blood vessels across the body (Murthy 2020).

Feelings of loneliness impede executive control and self-regulation, making us less kind to others (Twenge et al. 2007), more aggressive (Twenge et al. 2001), and more likely to engage in self-defeating behavior (Twenge et al. 2002). Research has also linked it to impulsive and selfish behavior (Cacioppo and Patrick 2008; Fumagalli et al. 2022).

Even more troubling is the fact that if loneliness is deemed harmful, it is apparently not as simple as just creating opportunities to interact with others.

For example, it has been reported that the negative effects of loneliness are greater when you are with someone than when you are alone (Stavrova and Ren 2023). Furthermore, it has also been shown that being with others does not necessarily increase the sense of well-being for those who feel strongly lonely. In fact, it has been suggested that it may even have the opposite effect. Many people may find that they feel even stronger loneliness when they are among many people.

The impact on children who have had to grow up in isolation, however, is even more severe (Blum 2002). It inhibits healthy development to a surprising degree, and the accounts of its effects are a truly harrowing read. Social disconnection can be more damaging to us humans, both children and adults, than we realize, and we'd do well to make this fact more widely known.[13]

For those interested in discussions on loneliness, consider reading books such as *Love at Goon Park: Harry Harlow and the Science of Affection* (Blum 2002), *Loneliness: Human Nature and the Need for Social Connection* (Cacioppo & Patrick 2008) and *Together: The Healing Power of Human Connection in a Sometimes Lonely World* (Murthy 2020). Additionally, for those intrigued by the power of connections, I encourage you to read *Friends: Understanding the Power of our Most Important Relationships* (Dunbar 2021).

[13] The global COVID-19 pandemic that descended upon us at the beginning of 2020 prompted lockdowns and the like all over the world, and more than a few people no doubt experienced a real sense of isolation as a consequence of these policies. An experiment conducted during the pandemic found that greater cooperative behavior in the ultimatum game was associated with more frequent contact with both friends and family during social isolation (Bland et al. 2022).

## 3.4 Dictator Game Variations

So far, we have discussed humans' extreme sensitivity to watching eyes and reputational concerns and, conversely, the tendency to pursue one's own self-interest when no one is watching.

We will now look at four variations on the dictator game that will help us dig a little deeper into the meaning of Player 1's behavior.[14]

**Variation 1: Don't mind if I do!**

The first variation is a dictator game with the added option of taking something away from your opponent.

First, Player 1 and Player 2 each receive $5. On top of that, Player 1 receives another $5 and is asked to decide how to split this $5 with Player 2 in basic dictator game fashion. Then in subsequent runs of the game, the options available to Player 1 are changed slightly (List 2007; Zhang and Ortmann 2014). This study involved undergraduates at the University of Maryland; Experiment 1 had 24 participants, Experiment 2 had 46, and Experiment 3 had 50.

«Experiment 1» Dictator game, basic version (Fig. 3.16)

In the basic game, Player 1—the dictator—gave Player 2 $1.33 out of the $5. The average fraction of the stake offered was around 25%, and about 70% of people gave some amount of money to their opponent. Around 30% of people in the Player 1 role kept everything to themselves and offered nothing to the other person. All of these outcomes were pretty standard and in accord with previous dictator game experiments. The amounts appearing below the illustrations represents the amount that each player ultimately received.

As Player 1:

Around 70% of people gave at least something to Player 2
Around 30% of people gave nothing to Player 2

[14] To simplify the discussion of this area of research, I've tweaked some of the monetary amounts to make them nice round numbers.

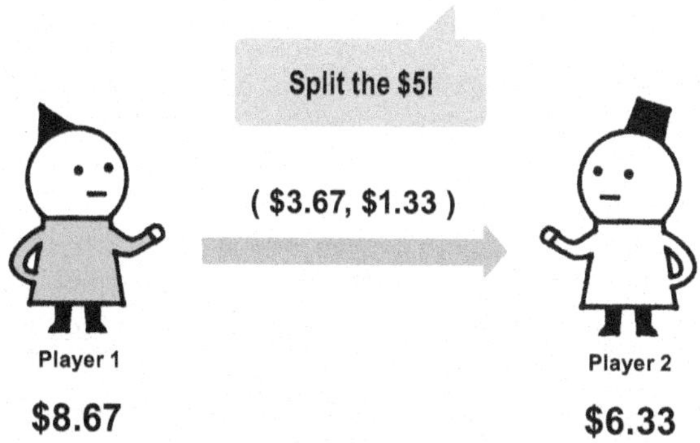


**Fig. 3.16** Dictator game, basic version

«Experiment 2» Option to take up to $1 from Player 2 (Fig. 3.17)

Here, Player 1 is again free to split the $5 with Player 2 in any manner but now also has the option of taking up to $1 from Player 2.

Given this option, the amount Player 1 gave to Player 2 plummeted to $0.33! Not only did the proportion of people who gave nothing to Player 2 rise to roughly 45%, but some 20% actually took the full $1 from Player 2!

As Player 1:

Around 35% of people gave at least something to Player 2
Around 45% of people neither gave nor took anything from Player 2
Around 20% of people took $1 from Player 2

**Fig. 3.17** Dictator game, option to take up to $1

«Experiment 3» Option to take up to $5 from Player 2 (Fig. 3.18)

In this final case, Player 1 may, as usual, split the $5 with Player 2, and now also has the option of taking up to $5 away from Player 2.

And given this option, Player 1 on average took around $2.48 from Player 2. That's right, "took", not "gave"! Hardly anyone gave anything to Player 2, around 30% gave nothing, and around 60% took money away. And some 40% of those purloined the entire $5! It's almost as if Player 1 had started tightening the thumbscrews on Player 2. What happened to all that beautiful altruism and concern for others' wellbeing?

As Player 1:

Around 10% gave at least something to Player 2
Around 30% of people neither gave nor took anything from Player 2
Around 60% of people took money away from Player 2
(of which, over[says "around" above] 40% took the entire $5)

**Fig. 3.18** Dictator game, option to take up to $5

Figure 3.19 summarizes the results of Experiments 1 to 3. This quite clearly shows that as Player 1 is given the option to take more and more from Player 2, the average amount actually given to Player 2 falls and, ultimately, Player 1 ends up taking money away from Player 2.

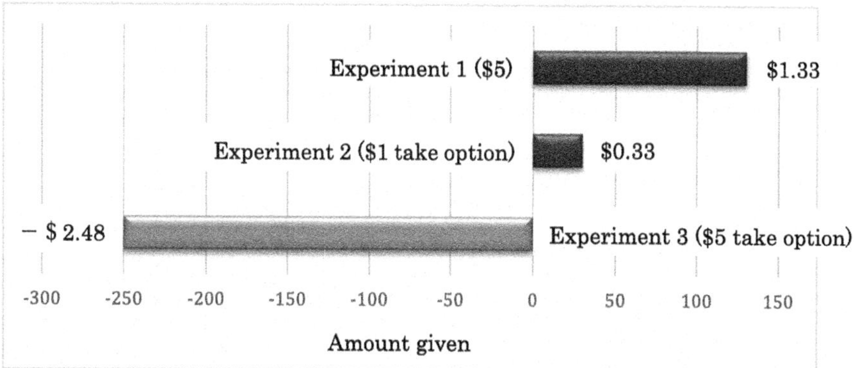


**Fig. 3.19** Average amount given to player 2 in each experiment. *Source* Prepared by the author based on List 2007 (with slight wording changes)

Once the option was presented, Player 1 began taking money away from Player 2. Although Player 1 did not (on average) squeeze everything out of Player 2, you'd be hard pressed to call this sort of behavior altruistic. In avoiding the worst option and not taking all of the money, is Player 1 perhaps exhibiting some concern for watching eyes and reputational effects and, in effect, saying, "I'm not *that* greedy!"? (List himself is said to have insisted: "It's certainly not altruism we've been seeing").

A slight change in the game's parameters produces a large change in the amount shared with Player 2. A little nudge is all that is required to shift the standard by which we deem certain behavior to be OK, or in other words, our answer to the question "How much can I get away with?" or "How much can I take before I've gone too far?".[15]

**Variation 2: What I don't know won't hurt me (Fig. 3.20)**

Next, we'll consider a dictator game in which Player 1 does not initially know Player 2's payoff (Dana et al. 2007). This study involved undergraduates at the University of Pittsburgh; Experiment 1 had 38 participants (19 pairs) and Experiment 2 had 64 (32 pairs).

A simple version of the dictator game was used in which the player is presented with only two choices. A computer screen displays two options for splitting the stake between Player 1 and Player 2. Player 1 is simply asked to choose one of these options.

[15] A similar study conducted in the African nation of Botswana found that although the average amount Player 1 gave to Player 2 was smaller than in Western societies (so-called WEIRD societies, as discussed in the brief aside below), the sequential drops in the amount given were much smaller (Drouvelis 2023).

«Experiment 1» Basic game, two choices

In Experiment 1, Player 1 was given the following two choices.

| Payoffs | P1 | P2 |
|---|---|---|
| Option 1: | $5 | $5 |
| Option 2: | $6 | $1 |

Option 1 splits the money evenly. Player 1 and Player 2 receive $5 each. Option 2 benefits Player 1 but has Player 2 losing out heavily. When asked to choose between these two options, over 70% of people in Player 1's seat chose Option 1 (equal split) despite knowing that they would be losing out (passing up the opportunity for an extra $1). Is this perhaps evidence for altruism to strangers after all?

«Experiment 2» Player 2's payoff initially hidden

In Experiment 2, Player 2's payoff was hidden from Player 1.

As before, Player 1 knows only that Player 2's payoff will either be $1 or $5. Player 1 does not know, however, which option will provide which payoff to Player 2. On screen, Player 2's payoff appeared as a question mark.

| Payoffs | P1 | P2 |
|---|---|---|
| Option 1: | $5 | ? |
| Option 2: | $6 | ? |

However, Player 1 had the ability to reveal Player 2's payoffs simply by clicking a button on the screen. Now, clicking a button does not take a whole lot of effort, so if Player 1 were going to spare a thought for the anonymous stranger on the other end (Player 2) in determining how to split the stake, you would expect him to reveal the payoffs and make his decision on that basis.

Yet, despite knowing how easy it would be to reveal the payoffs, roughly half of the Player 1 cohort did not even bother to look at what Player 2 would be receiving. And over 60% of them chose Option 2, which was clearly beneficial for them!

These players could honestly say, if asked afterwards, that they didn't know what effect choosing the $6 option would have on the other player (Kurzban 2011). To be sure, that would not be a lie. But the more honest answer would be to admit not having bothered to find out, despite it being easy to do so. This behavior is known as **strategic ignorance**.[16]

"I had no idea, so what else could I do?" rationalizes Player 1. And so again we find that a slight change to the game has made consideration for an anonymous stranger evaporate (Fig. 3.20).

**Fig. 3.20** Player 2's payoff hidden

**Variation 3: What you don't know won't hurt me (Fig. 3.21)**

In the third variation we consider, Player 1 is able to take an amount of money home without this choice ever being communicated to Player 2 (Dana et al. 2006). The participants in this study were 80 students (40 dictators) from Carnegie Mellon University.

Here, before deciding how to divide the $10 received from the researcher, Player 1 is given the option of paying $1 to the researcher to ensure Player 2 never knows there was a plan to split any money at all, allowing Player 1 to quietly take the remaining $9 home. That is, Player 1 is given an option to "exit" the game at a cost of $1 without Player 2 ever knowing that a game was afoot.

[16] Subsequent research has shown that even if remaining ignorant carries a cost, people will still choose to pay the cost and remain strategically ignorant, and moreover, that people who are willing to pay the cost to remain ignorant tend to choose options that result in a greater payoff to themselves (Grossman and van der Weele 2017).

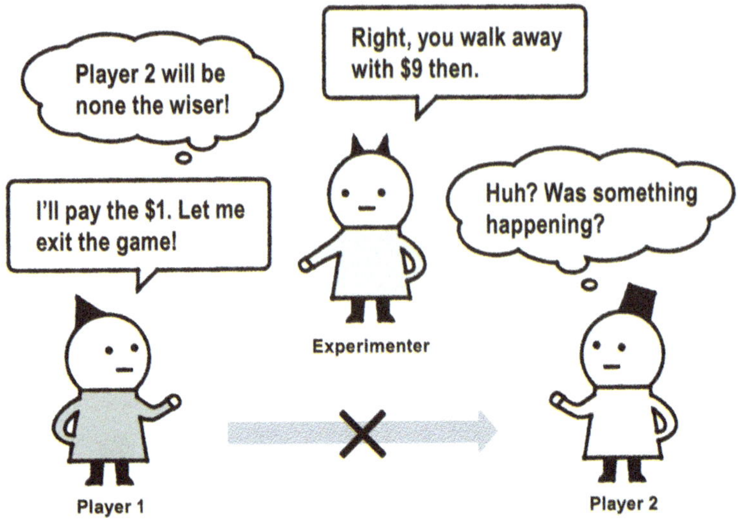


**Fig. 3.21** Game exit option

Roughly 30% of the participants chose to pay the researcher $1 to exit the game. In a subsequent replication study, around 40% of participants chose this exit option. Now in this case, Player 1 receives $9, so instead of paying $1 to exit the game, he could have chosen a ($9, $1) split, meaning that his gain would remain the same but Player 2 would also receive $1. So if Player 1 were altruistic and concerned about others' welfare, he could have given the $1 to Player 2 instead of wasting it. On the other hand, if he were simply driven by a desire for money, he could have played the game and chosen a ($10, $0) split, increasing his payoff by $1.

Yet in reality, Player 1 chose neither of those paths, instead opting to pay for the exit option, thereby "wasting" $1, so that Player 2 would not find out about his "greedy" choice. Replication work has produced similar results in Japan (Hashimoto et al. 2014). Based on these findings, we can almost hear Player 1 saying something like, "I do want the money, but I don't want the other person to think I'm so greedy!".[17]

[17] Some readers might actually feel that the 30–40% prevalence of the exit option among the Player 1 cohort is rather small. The size of this figure is possibly due to a fear that the researcher will regard the player not only as an unfair and self-interested opportunist but also as an abject coward lacking the decency to at least let the other player know the score.

**Status quo bias** may also have been at play. Status quo bias in this context is the tendency for people to be unwilling to change a choice once made—the tendency to want to maintain the status quo, in other words. In the study discussed in this chapter, Player 1 is only informed of the exit option after first thinking through the decision of how to divide the $10. Our underlying assumption is that more people might choose the exit option if told about it from the outset.

Indeed, subsequent work has shown that people are willing to pay up to 20% of the stake if they can exit without Player 2 knowing they have made the greedy choice (Broberg et al. 2007). In terms of the examples presented here, this means people are willing to pay up to $2 to exit the game. The researchers showed that when the cost of exiting is 10% of the stake, or $1 in terms of the examples in this chapter, about 60% of players choose to exit.

**Variation 4: Who's the fairest of them all?**

The final study looks at what happens when people don't know who made the proposal (Andreoni and Bernheim 2009). The experimental subjects were 120 volunteers from undergraduate economic courses at the University of Wisconsin–Madison.

This is how it was set up. When Player 1 proposes a split, a computer program sometimes overrides that proposal in favor of a predetermined split proposal. Say, for instance, that Player 1 proposed a 50–50 split. Sometimes the program will simply veto that and put forward a different proposal.

And the computer program's proposals were quite advantageous to Player 1: ($9.50, $0.50), ($10, $0), and so forth. Hence, when highly unfair proposals like this were made, Player 2 had no idea whether the proposal had actually come from Player 1 or from the computer program by fiat. Both Player 1 and Player 2 were well aware of this in advance (Fig. 3.22).

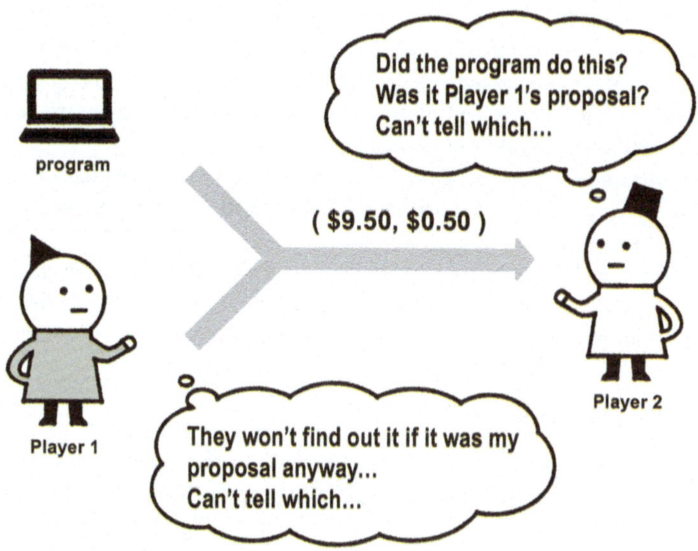


**Fig. 3.22** Player 1 can hide behind the computer program's proposals

Now, the more likely it became for the computer program to make suggestions advantageous to Player 1, the more Player 1 himself began to make the same proposals as the program. So once it was evident that Player 2 would be unable to tell whether Player 1 had made a particular proposal or not, Player 1 increasingly put forward unfair proposals for his own benefit.

We can almost hear Player 1 say, "Well, it's not like Player 2 can tell the difference anyway!" In the words of the researchers, Andreoni and Bernheim, "people like to be perceived as fair" rather than actually be fair.

However, this doesn't necessarily mean that there is no hope for humanity. Ultimatum game experiments with an exit option have shown that people tend to shy away from situations where they can deceive others (Shalvi et al. 2011). Despite the temptation to cheat in situations where it's possible, this gives us evidence of our inclination to steer clear of such circumstances altogether. And this is another aspect of our behavior that should pique our interest in human nature.

**We care about *looking* good**

Okay, so we've looked at four variations of the experiment. Time for another quick recap. Experiment 1 showed that people actually start taking from others once they are told it is okay to do so. Experiment 2 showed that when the opponent's payoff is hidden, people do not even bother to reveal it regardless of how easy it would be to do so, and simply choose the option that gives themselves the higher payoff. Experiment 3 showed that people are happy to pay a cost to prevent others from knowing that they have taken the greedy path. And finally, Experiment 4 showed that if a computer program provides options that are advantageous to the player, and if there is a way of "hiding" behind the program's proposals, the player will take advantage of this to choose options that unfairly benefit himself. Similar findings have also been teased out of different games (Katzir et al. 2021).

A look through the results of these variations suggests that Player 1's behavior, which initially might have (ostensibly) demonstrated concern for others' welfare, actually tells us something quite different.

"We care more about looking good than about truly being good," wrote prominent social psychologist Jonathan Haidt (Haidt 2012, p. 220). And the studies we have looked at do indeed suggest that this is part of our psyche, however regrettable that may be.

**A brief aside: Why so WEIRDy?**

It is crucial to note that most of the participants in the experiments discussed in this book were predominantly from specific regions such as the United States and Europe.

Henrich and colleagues have pushed back on evidence from behavioral experiments predominantly involving participants from WEIRD cultures (Henrich et al. 2010b). WEIRD, as you may know, refers to Western, educated, industrialized, rich, and democratic. For instance, in the context of adolescent developmental neuroscience, 99% of the samples used in the 80 publications covered by five recent meta-analyses were derived from Western countries (Qu et al. 2021). According to an extensive investigation led by Henrich and his colleagues, participants in experiments related to human psychology and behavior were found to be 96% from Northern Europe, North America, or Australia. Among them, approximately 70% were reported to be American university students (Henrich 2020).

Individuals from WEIRD cultures, characterized by high individualism, low conformity to group norms, heightened self-focus, the valuing of self-restraint, patience, excellence in analytical thinking, and a high level of trust towards strangers, display distinct features when compared to people from other regions or the majority of those who have lived in the past (Henrich 2020). (Those interested in how WEIRD people are in fact weird should definitely check out Heinrich's own work, *The WEIRDest People in the World* (Henrich 2020)).

However, the problem goes beyond just that. It is also imperative to be aware that participants in the experiments presented in this book are not only from such regions but also frequently students from universities in those areas, often from so-called prestigious universities. In one dictator game meta-study, for instance, 94.7% of participants covered were students (Lim and Yu 2015).

Research has shown that, in general, students tend to be more selfish than the general populace in games like the dictator game (Belot et al. 2015, Lim & Yu 2015). Others have reported that adults over about the age of 25 in Western societies have a strong tendency to divide the stake evenly (Henrich 2017), and that in experiments involving non-student participants in Asia, older players tended to give more (Ogawa et al. 2020).[18]

The World Health Organization (WHO) defines adolescence as the period between the ages of 10 and 19 inclusive (https://www.who.int/health-topics/adolescent-health/#tab=tab_1), but recent years have been witness to debate about pushing back the upper bound to define the range as, say, from 10 to 24 years (Sawyer et al. 2018). We know that during this period called adolescence, people put great emphasis on their bonds with peers and can be prone to choosing risky and even reckless courses of action. We now think the human brain does not actually finish developing until people's late 20s or even around 30 years of age. This seems to be when the prefrontal cortex, the seat of higher cognitive functions that endows us with human reasoning and creativity, fully matures (myelination is complete).

That the bulk of the experimental data we have concerns university students means that the bulk of the data we have concerns people in the throes of adolescence. Adding "young" to WEIRD to make it WEIRDy, I have to say that we must always keep in mind, at this point, that it is dangerous to draw conclusions about humans in general based on WEIRDy results alone.[19]

[18] The age effect is apparently not limited to the act of sharing with strangers. General prosociality also appears to increase with age (Matsumoto et al. 2016; Cutler et al. 2021).

[19] This proviso, of course, also applies to this book. That is to say, since many of the studies cited here are also based on WEIRDy data, appropriate caution must also be exercised when interpreting the conclusions of this publication. It's truly unfortunate that, due to space limitations, almost all discussions on diversity had to be omitted from this book. However, I firmly believe that advancing such research is indispensable for exploring these issues.

**Summary**

About the proposer (Player 1)

- Increases proposed amounts to a degree to avoid being told "no".
- Tries to avoid being seen in a negative light in the presence of watching eyes, pursues own self-interests when no one is watching.
- The less he senses he is being watched, the stronger his tendency to pursue his own self-interests.
- Watching eyes need not be real ones.
- Positive feedback produces a real sense of joy in the brain, isolation and ostracism produce a pain response.
- If told he may take something away from Player 2, he will, but he won't take the entire amount.
- If permitted to remain ignorant of Player 2's payoff, he will choose strategic ignorance and put priority on his own interests.
- He will happily pay a cost to prevent Player 2 from finding out that he has acted selfishly.
- If Player 2 will be unable to tell if he made the proposal or not, he will put forward proposals that are pretty unfair.

To summarize the discussion thus far, one thing Player 1 is concerned about, in addition to the strategic matter of sharing just enough to avoid proposals being rejected by Player 2, is watching eyes and reputation—that is, what others think of him. While we humans are so concerned about being seen that we even react unconsciously to what might be perceived as watching eyes, in truth, we really care more about looking good than about truly being good, and about being able to tell ourselves that we are good.

*So convenient a thing it is to be a reasonable creature, since it enables one to find or make a reason for everything one has a mind to do.*
*—Benjamin Franklin, The Autobiography of Benjamin Franklin, c.1791*

## Chapter 3 Supplement

### On the effect of watching eyes

In Sect. 3.2, we discussed the possibility that humans react quite sensitively to non-real watching eyes, such eyespots on a poster. It must be noted, however, that this watching-eyes effect seems to be much more limited than initially thought (Kobayashi 2024).

Researchers have reported, for instance, that although the watching-eyes effect appears when few people are present at the location of the experiment, the effect weakens when many people are present (Ernest-Jones et al. 2011; Powell et al. 2012; Ekström 2012; Oda and Ichihashi 2016; Oda 2019). When lots of people are around, there are already lots of actual watching eyes to be concerned about, so fake eyes do not produce much of an "incremental" effect.

In social psychology terms, the group to which you belong is called your **in-group**, and groups to which you do not belong are called your **out-groups**. Researchers have found that the watching-eyes effect can produce favoritism toward in-group members, but that the effect is weak when it comes to out-group members (Mifune et al. 2010). Hence, the watching-eyes effect seems to be weak when we are dealing with people with whom we are unlikely to interact again in the future. As the saying goes, in one form or another, *what happens on vacation stays on vacation.*

The effect also seems to be weak if the eyes are not looking directly at you but are instead closed or averted (Manesi et al. 2016), and the effect seems to disappear when subjects are placed in a dark, sound-proof room (Tane and Takezawa 2011). Researchers have also found the effect to be weak for other facial features, which, like eyes, could also indicate the presence of another person, and that eyes produce the strongest effect (Kelsey et al. 2018). Others have reported no real effect when people are exposed to cues designed to remind them of their peers (Baillon et al. 2013), when auditory cues of the presence of others are eliminated (Haley and Fessler 2005), and when other people are present in the same room but anonymity is assured (Lamba and Mace 2010). The effect seems to be weak when someone else is merely nearby, and it's quite likely that people need to have a clear sense that the person is actually watching them for the watching-eyes effect to manifest itself (Vaish et al. 2017). These results accord with what we might naturally expect if the effect of eyes is driven by people's reputational concerns.

A study following up on the coffee room experiment discussed earlier found that when two large flowers were positioned side by side on the banner in a manner compositionally similar to the photographs of eyes, the difference in effect disappeared (Carbon and Hesslinger 2011). This also seems quite reasonable if we think back to the experiments that used three-dot eyespots.

The watching-eyes effect is also highly transitory. The effect is present right after people have seen the watching eyes but disappears after some period of exposure (Sparks & Barclay 2013). It is widely known that humans generally become accustomed to stimuli over time. And so the finding that we might quickly become accustomed to watching-eyes stimuli seems quite reasonable. Researchers have also argued that the existence of social norms around the sort of behavior that is expected of people prevailing in the location where an experiment is conducted will, of course, influence behavior (Kawamura and Kusumi 2017). Yet others have found that while watching eyes alone may not be effective in some situations, an effect may be elicited depending on the manner in which the watching eyes are presented (Ayal et al. 2021).

In fact, an increasing number of studies are finding no watching-eyes effect (including Fehr and Schneider 2009; Cai et al. 2015; Matsugasaki et al. 2015; Lennon et al. 2017; Manesi and Pollet 2017; Koornneef et al. 2018; and Rotella et al. 2021), and a couple of fairly recent meta-analyses found no watching-eyes effect either (Northover et al. 2017a; Northover et al. 2017b). I think it's safe to say that the matter is not as simple as just putting up a picture of watching eyes wherever and whenever and expecting people's behavior to magically incline toward prosociality.

Taking all of these studies in, it seems to me that the question of whether (or not) such cognitive biases and behavioral anomalies exist does not lend itself to a binary "yes or no" answer (Kobayashi 2024).

I think the broader discussion about whether watching-eyes effects exist or not has so far been largely binary in nature. Yet the truth is that human cognitive tendencies exist on a continuous spectrum—they show up more strongly in some situations and more weakly in others. Multiple factors influence the conditions under which the watching-eyes effect manifests itself, including the number of people present at the location of the experiment, the direction in which the eyes are looking, the brightness (or darkness) of the location, the attributes of experimental subjects, and the length of exposure to stimuli. Similarly, multiple factors probably influence the strength or weakness with which human cognitive biases present themselves.

You may be interested to know that the word "nudge" carries a specific meaning in behavioral economics. Richard Thaler, winner of the 2017 the Nobel award in economics, defines a nudge as "any aspect of the choice architecture that alters people's behavior in a predictable way without forbidding any options or significantly changing their economic incentives" (Thaler 2008, p. 6). In other words, it could be described as a mechanism that helps people voluntarily make better choices for themselves, without explicitly telling them what to do or what not to do, and with little regard for profit or loss.

However, as the idea of a nudge in this sense become more widely known, we often hear people saying that it's not as effective as they thought it might have been. This may be partly because of the dualistic approach of looking at something as either working or not working, without consideration for these boundary conditions.

After all, could it not be said that at least some of the studies claiming to have failed to reproduce the watching-eyes effect have shed light on its boundary conditions? While brain research has shown that a direct gaze or eye contact can influence various aspects of social cognition and behavior, researchers have also pointed out that the way in which these effects manifest themselves is by no means simple (Senju & Johnson 2009). When and how the watching-eyes effect might (or might not) show up is a matter that still requires much careful consideration and study.

# Chapter 4
# We Hate Unfairness

## Why are we willing to pay to spurn?

## 4.1 Exploit Me at Your Peril!

**Player 2's predicament**

From the discussion so far, we know that Player 1 seems to be concerned about how they are perceived by others, or in other words, their own reputation. Let's now turn our attention to Player 2, the responder in the ultimatum game, and examine the second problem we identified: Why does Player 2 reject some offers at the cost of losing something of value?

Approached intuitively, this behavior may seem so obvious that it's difficult to tell what the problem is. Let's quickly recap how the ultimatum game works. Player 1 is to decide how to split a sum of money (we take $10 as the typical case), and Player 2 is to either agree or disagree with the proposed split. If Player 2 says, "Yes", the money will be divided accordingly; if she says, "No", the $10 is forfeit and neither player walks away with anything (Fig. 4.1).

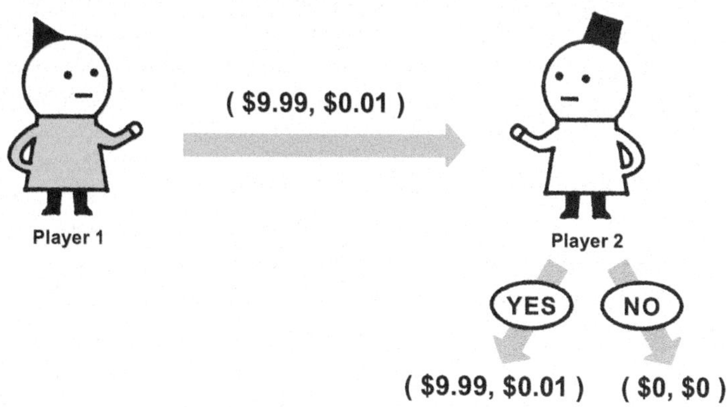


**Fig. 4.1** The ultimatum game: basic version


K. Kobayashi, *The Mystery of the Ultimatum Game*,
https://doi.org/10.1007/978-981-97-8979-5_4

Can you remember how Player 2 would respond if she were an Econ? Money is money. Every extra cent counts, so as long as she stands to receive at least one cent, she agrees to all proposals. The problem, however, is that, intuitively, this behavior just doesn't feel right!

This problem is very strange if you think carefully about it. It's not that Player 2 doesn't understand the situation, nor that she doesn't know what the (economically) most beneficial course of action is. She understands how the game works and what the optimal answer is, and yet, in real life, she rejects some proposals even at the cost of losing out on a share of the pie. This is possibly an even more difficult problem than the Player 1 conundrum.

When I ask my students, though, the answer just rolls off their collective tongues. "Because it's unfair!" Again, intuitively, this is quite a convincing answer, but we should delve a little deeper.

## It just isn't fair!

This brings us to the, in a sense, simple notion that people dislike unfair outcomes, an idea called **inequity aversion** (Fehr & Schmidt 1999). Inequity aversion says that because we humans abhor unfairness or inequity—when your opponent benefits more than you do—we will at times pay a cost to take action that cancels out the deficit. This is precisely what Player 2 is doing when she gives up money in the hand to reject a proposal from Player 1.

This is supported by neuroscience demonstrating that our brains dislike unfairness (Tricomi et al. 2010).

As discussed, when you receive a reward, like money, your brain's reward system, the part of the brain that responds to joy and pleasure, lights up. This is why it naturally feels good to get paid! Yet the brain's reward system does not only respond when you are the one receiving some sort of benefit or reward. The reward system (striatum) also lights up when the gap between you and your opponent is reduced, making a given situation fairer. And surprisingly, this same joy response is observed in the brain even when you have an advantage over your opponent, and your opponent, who is of course at a disadvantage to you, receives money, for instance, thus reducing the gap between the two of you. Joy, as far as the brain is concerned, lies in the closing of gaps, which can mean improving the situation of someone who is worse off than you.[1]

## Dealing with the old brain—unfairness as a natural enemy

Researchers have also shown that when a large disparity exists between experimental subjects, the **amygdala**—a part of the brain that responds when we feel fear or anxiety—lights up in prosocial subjects (Haruno and Frith 2010).

---

[1] When you learn that the brain registers joy when the situation of others worse off than you improves and disparities are thus reduced, you might be tempted to see this as a reflection of the beauty of the human soul. But not so fast. Flip it around, and what this means is that we rejoice in the misfortune of those who are better off than us. That is, the misfortune of others tastes so sweet! We have a word for this, by way of German, that you may be familiar with: "schadenfreude". For a deeper discussion, see the end-of-chapter supplement.

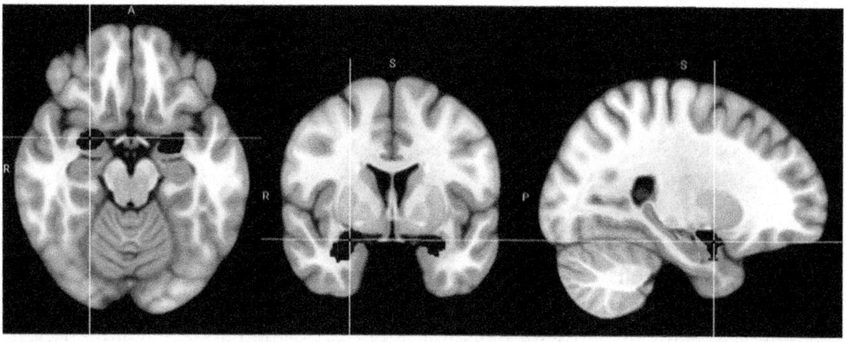


**Fig. 4.2** Amygdala (image provided by professor Asamizuya, created using MRIcroGL)

The amygdala is an almond-shaped structure located toward the back of the brain (Fig. 4.2). It is common to many animals and is considered part of the *old brain* in evolutionary terms. Sometimes called the brain's alarm bell or warning system, it reacts surprisingly quickly when danger is detected, sending out signals to prepare the body for fight or flight, and hence it is thought to be what originally controlled our response to natural enemies and predators. The evolutionarily new brain, or neocortex, meanwhile, is responsible for our higher cognitive functions, such as the reasoning and creativity that marks us as human. This is located behind the forehead in what is called the prefrontal area (Watanabe 2013).

People quite naturally thought that the new brain, our center of higher reasoning and decision-making, was responsible for singularly human patterns of thought such as the notion of fairness. In humans, went the reasoning, the new brain would regulate the old brain's natural, animalistic impulses. "I want more," says the old brain, to which the new brain admonishes, "Think of others first!".

Research, however, has proved otherwise. It is in fact the amygdala, the part of the brain that responds to enemies and predators, that responds to unfairness. This result—that the old brain is involved in decisions about fairness, something people had assumed was unique to the human intellect—was received with no small amount of surprise. Indeed, I remember my own astonishment upon discovering that the amygdala was involved in our sense of fairness. A trove of subsequent research supports this finding. This perhaps suggests that the tendency to dislike unfairness is deep-rooted in the brain.[2]

**Infants and babies "spit the dummy" over unfairness**

We have also discovered that this tendency to dislike unfairness manifests from a very young age.

Distaste for unfairness can be divided broadly into two types: disadvantageous inequity aversion—dislike for a situation in which you have less than someone else—and advantageous inequity aversion—dislike for a situation in which you have more than someone else.

---

[2] Researchers have also reported a relationship between greater pain sensitivity and the experience of unfairness (Wang et al. 2019). They argue that both pain and the sense of unfairness are a kind of alarm system.

Researchers have shown that even young children around kindergarten or nursery school age have a tendency to try to divide resources equally (Sigelman and Waitzman 1992). Later on, largely by around the age of 4, children show an aversion to resource distributions that put themselves at a disadvantage, and then by around 7 or 8, children tend to avoid distributions that put themselves at an unfair advantage, preferring to reduce their own share of resources when they are allocated excessively more than someone else (Blake and MacAuliffe 2011; Fehr et al. 2008, Shaw and Olson 2012; Sutter et al. 2019; Engelmann and Tomasello 2019). Also around the age of 7 or 8, children have a tendency to allocate more to others when the pot consists of an odd number of items and thus cannot be divided equally (Shaw et al. 2016). And children as young as 5 are satisfied with receiving less than someone else as long as a fair procedure, like rock paper scissors, has been followed to make the allocation (Shaw and Olson 2014; Grocke et al. 2015; Engelmann and Tomasello 2019). Researchers have also found that 3-year-olds are able to take merit (as in who contributed more) into account when distributing goods, and that even 21-month-olds expect resources to be divided according to effort (Sloane et al. 2012). So it seems that we humans develop a complex concept of fairness that goes beyond mere equality, or equal distribution of resources, during our early years.

Some research even suggests that children as young as 15 months (Sommerville et al. 2013), 9 months, and 4 months (Dawkins et al. 2019) may already see unfair distributions of goods as dubious.[3] As Sutter et al. (2019; p. 113) note, "Fairness concerns seem deep rooted and early developed." (If you are interested in how on earth do they experiment with babies who are said to love fairness, please read *"Just Babies: The Origins of Good and Evil,"* (Bloom 2015). Anyone who works with small children has no doubt realized how surprisingly common it is for our young ones to rail against perceived injustice with expressions like "That's not fair!" Come to think of it, the phrase 'that's not fair' is still frequently used by our kids at home. This may indeed have something to do with human sensitivity to unfairness and injustice as discussed here.

**Inequity aversion in the animal kingdom**

Perhaps even more surprising is the possibility that animals may also be averse to unfairness.

Suppose, for example, that you have two capuchin monkeys perform the same task, and then provide unequal rewards; that is, you give one a more attractive reward than the other (Brosnan and De Waal 2003). To one, you give some cucumber, a rather pedestrian reward, and to the other, you give a grape, a highly prized food. While capuchins love grapes, they apparently are not wont to turn down some cucumber either, and despite this, when the researchers performed this experiment, the capuchin who received cucumber would get rather mad and even throw the cucumber back

[3] In the case of children younger than 1, however, this only holds under very simple conditions, and it is not always consistent. For details, see Dawkins et al. (2019). Furthermore, as children grow older, they generally behave more fairly, but from around school-age, their perception of what constitutes fair starts to be heavily influenced by the prevailing norms of their respective societies (Tomasello 2019).

at the researcher in protest. You can see footage of this in a TED Talk given by the prominent animal behaviorist Frans de Waal. The sight of a capuchin monkey desperately fixated on receiving the better reward, twisting and squirming as if to say, "No, there's been some mistake! You've given me the wrong one!" is absolutely adorable and well worth a watch.[4]

Based on similar experiments in other animals, Brosnan (2013) argues that among the primates, chimpanzees, bonobos, macaques, and capuchin monkeys reject unfairness, as indicated in Fig. 4.3.[5] Other researchers have reported that marmosets also display unfairness aversion when the task is set up differently (Yasue et al. 2018).

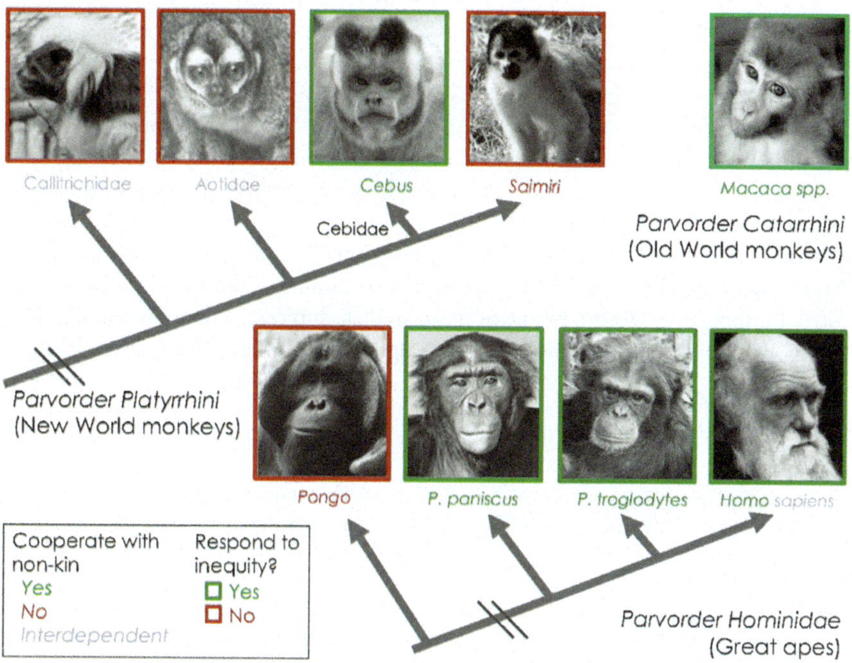


**Fig. 4.3** Summary of experimental evidence on reactions to the unfair distribution of resources. *Source* Brosnan (2013)

[4] De Waal's TED talk is titled "Moral behavior in animals". In addition to the capuchin monkey experiment discussed here, the video also contains footage of experiments in other animals and is a lot of fun to watch (https://www.ted.com/talks/frans_de_waal_do_animals_have_morals).

[5] What's intriguing in Fig. 4.3 is that it's not necessarily the primates closest to humans on the phylogenetic tree that are averse to unfairness. Orangutans, for instance, are known for their intelligence and are closer to humans on the tree than macaques or capuchin monkeys, yet they seem to tolerate unfairness.

Orangutans live alone in the wild, whereas those species that seem to reject unfairness are known to form close cooperative relationships, which includes alliances and partnerships with familially unrelated individuals. Brosnan argues that this difference may reflect the differing social and ecological pressures faced by each species. A lack of concern for fairness with respect to other individuals may make it difficult to form cooperative relationships.

Subsequent research indicates that individuals with more social personalities are more sensitive to unfairness (Brosnan et al. 2015).

Researchers have also observed similar behavior in dogs (Range et al. 2009; Range et al. 2012; Brucks et al. 2016; see McGetrick and Range 2018 for an overview), wolves (Essler et al. 2017), rats (Oberliessen et al. 2016), and some birds such as crows and ravens (Wascher and Bugnyar 2013), and have reported that chimpanzees may even refuse a reward when they receive a better one than their partner (Brosnan et al. 2010).[6] In view of these findings, De Waal (2014, p. 55) notes, "Survival depends on sharing, which explains why both humans and animals are exquisitely sensitive to fair divisions."[7]

**How we react in the ultimatum game**

So how does the human brain, which is "exquisitely sensitive" to fairness, react to ultimatum game proposals?

Let's start by considering fair proposals. Now, Player 1's proposal, by itself, is not enough to determine fairness in the ultimatum game. Say Player 1 proposes to give $5 to Player 2. The question of whether this will seem fair or not hinges very much on whether Player 1 was initially endowed with something like $10 or something like $1,000.

Studies have shown that the brain's reward system (striatum) responds differently depending on whether we are treated fairly or unfairly (Tabibnia and Lieberman 2007; Tabibnia et al. 2008). In one of these studies, subjects' self-reported happiness ratings were higher in response to fair, as opposed to unfair, offers, even when the

[6] In the case of dogs, however, unlike primates, it was reported that responses did not differ with the quality of reward (Range et al. 2009); i.e., dogs would stop working if they saw another dog receiving food while they did not, but they didn't seem to care if their own reward was bread while the other dog's was a sausage.

[7] Some, however, have criticized the experimental procedures and interpretation of results relating to unfairness aversion in animals, and some researchers have also reported contradictory findings (Henrich 2004; Bräuer et al. 2006; McAuliffe et al. 2015b; Tomasello 2016; McGetrick and Friederike 2018). Oberliessen & Kalenscher (2019) present a brief overview in Table 1, summarizing both positive and negative evidence of inequity aversion across animal species. Notably, a recent meta-study published during this book's revision raises concerns, arguing that while animals may display "disappointment" in response to "poor" rewards, such reactions are likely not driven by comparisons with other individuals (Ritov et al. 2024). The trajectory of future research will need careful attention.

Furthermore, an early ultimatum game experiment with chimpanzees found that chimpanzees behave "rationally" (Jensen et al. 2007a); but arguing that this experiment used an overly complex mechanical apparatus, Proctor et al. (2013), using a different methodology, were able to demonstrate that chimpanzees do indeed behave fairly in a manner similar to humans, and this line of research is ongoing. For example, it was reported that in an ultimatum game experiment with bonobos and chimpanzees, when the receiver (Player 2) is the only one who benefits from a refusal, the receiver rejects selfish offers, and the proposer (Player 1) distributes more to the receiver. This suggests that both bonobos and chimpanzees can behave strategically according to the situation at hand (Sánchez-Amaro and Rossano 2021).

For a review from the standpoint that there is also inequity aversion in the animal world, see, for example, Sapolsky (2017), and for a critical review, see, for example, Tomasello (2019).

absolute monetary value of the offer was the same. Fairness and equity matter to us, beyond just raw monetary incentives. We like being treated fairly.[8]

**Reacting to unfair proposals**

How about small, unfair offers? Research indicates that such offers light up an area of the brain (anterior insula) thought to be responsible for strong negative emotions like disgust and anger (Sanfey et al. 2003) (As discussed in Sect. 3.3, the anterior insula is also involved in the processing of pain).

The more unfair the situation, the stronger the reaction in the anterior insula. The researchers also observed that people who had a stronger reaction in the anterior insula more frequently rejected low-ball proposals. Subsequent research backs up this finding that the anterior insula, responsible for disgust, anger, and the like, lights up in response to unfair offers (Gabay et al. 2014; Cheng et al. 2017). One study found that unfair offers also evoked the same sort of facial expression as looking at something that is physically disgusting (Chapman et al. 2009). Ultimately, it is probably quite reasonable to say that people harbor intense anger and disgust towards unfair proposals.

**Sending a message (or a parting shot)**

To further understand what it feels like to be presented with an unfair offer, an experiment was conducted in which Player 2 was allowed to send a message to Player 1 before deciding whether to accept or reject Player 1's proposal (Xiao and Houser 2005).

In the study, low-ball offers elicited fairly palpable outrage in some participants, with messages featuring phrases like: "Sorry, I'm a person too", "Not fair!", "Too selfish", "Dude, that's kinda greedy", and "You suck" (Xiao and Houser 2005; example messages are given in Table 2 of their paper) (Fig. 4.4).[9]

---

[8] There exists an intriguing study in which intoxicated individuals participated in the Ultimatum Game (Morewedge et al. 2014). When they were drunk and presumably had less self-control, they were more likely to reject unfair offers compared to when they were sober. It has also been reported that depleting self-control leads to a higher rejection rate of unfair offers (Halali et al. 2014). Conversely, when they were allowed more time before responding to offers, which may have given them an opportunity to cool down and regain self-control, the rejection rate of unfair offers dropped significantly from 70 to 20% (Grimm and Mengel 2011).

Interestingly, however, the amount of proposals by Player 1 did not change much even when they were drunk (Morewedge et al. 2014). Additionally, when their self-control was depleted, they tended to reciprocate trust shown by others more often (Halali et al. 2014). These results seem to suggest that humans do not simply behave selfishly when their self-control is diminished.

[9] Interestingly, though, once granted the ability to send negative messages like these, Player 2 apparently became more receptive to small offers.

At the same time as unfair proposals evoke intense disgust and anger, they probably also, and quite naturally, make you want to let the other player know how you really feel! A rejection of an unfair proposal, aside from its express purpose, also serves to spite the other player and communicate displeasure. Player 2's ability to fulfill that objective via a text message instead is thought to explain the observed increase in receptiveness to the proposals themselves. Allowing people to vent about unfair divisions perhaps helps to mollify them a little.

**Fig. 4.4** The ability to send a message

Post-experimental studies have also shown that people who reject small offers express anger at the proposer's greed and a desire to punish them for being unfair (Pillutla and Murnighan 1996). We humans seem to experience strong negative emotions when presented with unfair offers.[10]

**Unfairness and intent—if you did this deliberately...**

I am sure you can understand the feeling of intense disgust and anger at unfair proposals. Research also has shown, though, that if there's anything the brain despises more than unfairness itself, it's the actual intent to treat someone unfairly (Sanfey et al. 2003).

To show this, the researchers started with a familiar setup: Player 1 is to divide $10. Now, the study participants (the human ones, at least) were all put in the Player 2 chair, and all the proposals were made by a computer program. However, those human participants were split into two groups: a group that knew the proposals were simply being made by a computer, and a group that believed that the proposals were being made by another person.

[10] Researchers have also reported that even when the payoff profile is altered so that when Player 2 rejects Player 1's offer, Player 2 receives nothing but Player 1's payoff is not reduced, Player 2 will still reject small offers, albeit not as frequently (Yamagishi et al. 2009, Ma et al. 2012).

And players still rejected offers even when they knew that the fact of the rejection would not be communicated to Player 1 or anyone else. A rejection in this situation does not serve to punish Player 1 by reducing his payoff, nor does it help to communicate displeasure or spite, nor does it have any reputational value that might come from letting people know you're not the sort of person to put up with such contempt. That players will still sometimes reject unfair offers in this situation is thought to be a reflection of just how strong the emotion experienced is.

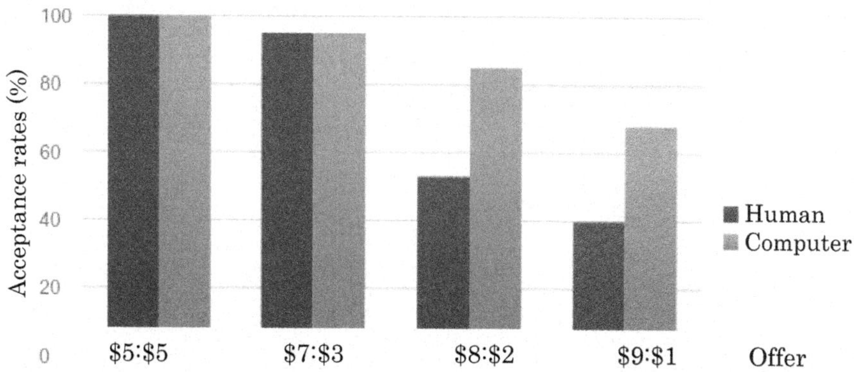


**Fig. 4.5** Differences in acceptance rates by proposed split (versus human and versus computer). *Source* Sanfey et al. (2003)

Look at Fig. 4.5. When people believed that the proposals were human-generated, the more unfair a proposal, the more likely it was to be rejected. But when the subjects knew that the proposals were computer-generated, they did not so readily reject proposals, even unfair ones. The problem, evidently, is not the proposed split itself, but the question of whether an unfair proposal was made by another human or not, which is to say, whether there is intent behind the proposal or not (Figs. 4.6 and 4.7).

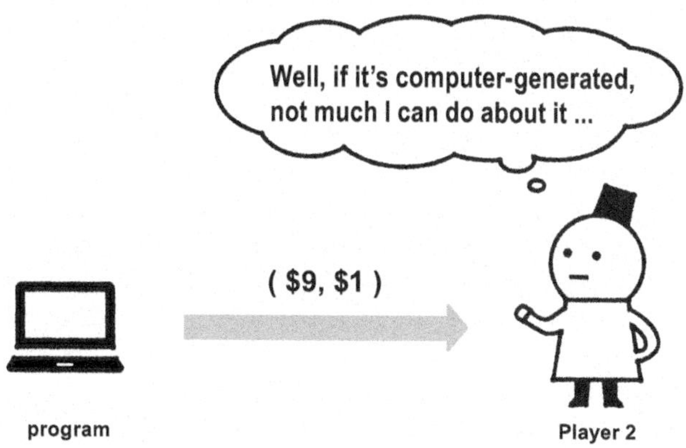


**Fig. 4.6** Proposal made by a computer

**Fig. 4.7** Proposal made by a human

When we looked at the brain's response to unfair offers, we found that the brain regions associated with emotions such as disgust and anger (anterior insula) responded more strongly to the same unfair proposal if they believed the proposal had been made by a human as opposed to when they knew it had been made by a program.

Many other studies have demonstrated the importance of intent. Unfair divisions are less likely to be rejected when they have been determined randomly (Blount 1995). And when a different set of alternatives is available to the proposer, an unfair division is less likely to be rejected if it is perceived as unavoidable because no fairer alternatives were available than if it is perceived as intentionally unfair because a fairer alternative existed (Falk et al. 2003).

Some researchers have reported that damage caused intentionally is perceived as more unforgivable (Fehr et al. 2010), and moreover that intended harms are seen as worse than unintended harms, even when the magnitude of the damage itself is the same (Ames and Fiske 2013). Additional research has shown that unintentional bad outcomes may be forgiven, whereas people will punish behavior that deliberately sought to harm an opponent even if it did not, in the end, yield a bad outcome (Gago 2021). A fairly recent study indicates that the more one believes that inequality is caused by humans and is deliberate, the more one evaluates it from a moral point of view and experiences moral outrage (Cervone et al. 2024).

Studies show that not only do young children and chimpanzees understand whether intent is present or not (Call and Tomasello 1998, 2008) but infants as young as 8 months old can also make that distinction (Hamlin 2013).[11] Other research shows that intent affects how well we remember others' actions. The reason why

[11] In his 2016 book, prominent comparative cognitive scientist Michael Tomasello writes, "we must stress again that the notion of distributive justice is not fundamentally about 'the stuff' but, rather, about being treated fairly and with respect" (Tomasello 2016, p. 117). As the experimental work covered in this book indicates, whether we perceive a given way of dividing resources to be fair or not hinges on more than just the resulting division itself.

someone did something—whether they accidentally did something bad or deliberately conspired to do something with ill intent—seems to be more memorable than the act itself, or what the person actually did (Mesoudi et al. 2006; Redhead and Dunbar 2013; Dunbar 2021).

Our tendency for magnanimity over accidental transgressions but vindictiveness over deliberate ones is, for better or worse, probably something you are quite familiar with in everyday life. In my case, this topic evokes memories about my second son when he was very young. He was a very strong-willed child, so he would get upset with his brother, two years his senior, when he interrupted or hindered his playtime. Yet he was such a gentle and patient older brother as well and never showed any anger when his little sister, who was just a little baby, did the same thing (or even worse!). That was probably because she was not doing it on purpose. It amazed me to learn that children as young as three or four are so sensitive to the presence or absence of intent. The question of whether something is deliberate or not is crucial to us from a very early age, it seems.

**Being seen as inferior is infuriating**

The discussion so far has illuminated how we despise intentional unfairness. In my view, however, the rejection of proposals in the ultimatum game reflects more than just this disdain for (intentional) unfairness.

One study showed that people with a strong tendency to reject offers in the ultimatum game are not necessarily the type of people who actively choose behaviors that benefit society as a whole (Yamagishi et al. 2012). When they examined the relationship between behavior in different games, the researchers found that choosing to reject offers in the ultimatum game did not always mean that the responder was taking the initiative to act fairly or seeking to increase others' benefit, and that, instead, rejectors had a strong tendency to pursue their own self-interest. Subsequent work has produced similar results (Hoeft and Mill 2017). Another study found that high-testosterone men tend to reject low offers in the ultimatum game (Burnham 2007).

In this light, a decision to reject a low offer can be interpreted as an expression of resentment at having been intentionally treated unfairly. In other words, the responder may feel that the proposer's decision to deliberately make a small offer in violation of the (unwritten) norm that resources should be divided equally means that the proposer regards her as inferior, and she thus rails against this. The idea is that this sort of behavior serves to inoculate the person against the possibility of someone treating them as inferior in a similar manner in the future (Kobayashi 2021).[12] In other words, the behavior can be interpreted as an attempt to protect the person's future self by ensuring that others don't think of them as a pushover who will put up with any old nonsense.

---

[12] There need not be only one motive for rejecting an offer, of course. See the end-of-chapter supplement for a deeper look.

On a different note, I encourage readers interested in the significance of inequity in different societies to read Wilkinson and Pickett's *The Inner Level: How More Equal Societies Reduce Stress, Restore Sanity and Improve Everyone's Well-Being (2019).*

## 4.2 Taking a Loss to Show Who's Boss!

### Altruistic punishment

So far, we've discussed how being treated unfairly evokes intense disdain and anger in us. In other words, it seems that we just can't stand it and we won't put up with it (Fig. 4.8).

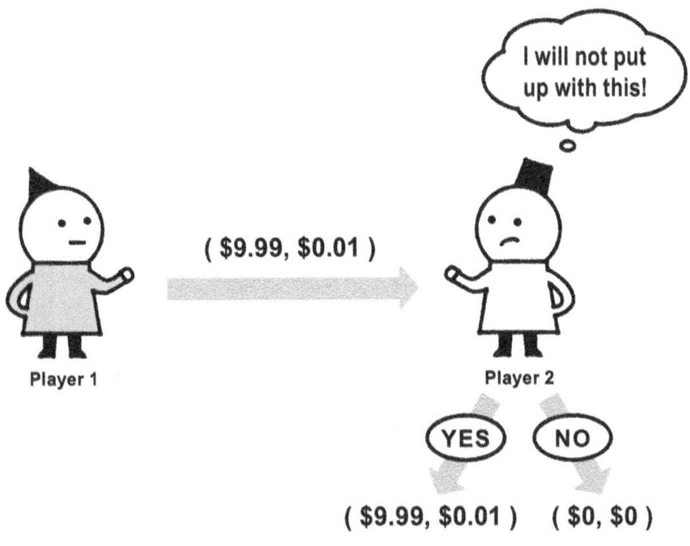


**Fig. 4.8** Intense disdain and anger at intentionally unfair treatment

This intense reaction is perhaps understandable, but even so, how can we explain why we are willing to take a loss in order to reject an unfair proposal? Why do we go so far as to reduce the payoff not only to our opponent but also to ourselves? This is the question we originally wanted to answer.

The act of punishing someone for a misdeed even at the cost of incurring a loss ourselves is called **altruistic punishment**. This is precisely what Player 2 is doing: taking a loss herself in order to punish a greedy opponent.

As a combination of the positive word "altruistic" and the negative word "punishment", the term may sound a little odd. To get our heads around it, let's look at an experiment conducted by renowned behavioral economist Ernst Fehr (Fehr and Gächter 2000; 2002).[13] The experiment involved 240 students.

First, the researchers put students into groups of a few people, and everyone was then asked to simultaneously choose whether they would behave cooperatively (in a prosocial manner) or uncooperatively (considering only the gain to themself). Group membership was completely reshuffled in each round of this exercise, and participants had no idea what current group members had chosen in the past. The cooperative behavior option was costly (involved taking a loss), so participants had an incentive to avoid it if possible. Hence, the setup was such that if everyone was an Econ, no one would select the cooperative option (Fig. 4.9).

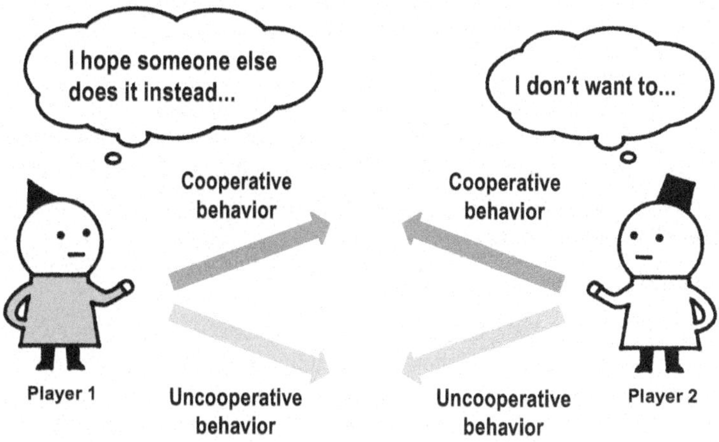


**Fig. 4.9** Choosing between cooperative and uncooperative behavior

This is where altruistic punishment is added to the mix. If a participant is in some way dissatisfied with how another group member is behaving—say they are being abjectly selfish—then, by paying a cost, the dissatisfied participant is able to punish said miscreant by confiscating money from them. Note here that a person who somehow benefits from the work of, or at the expense of, others without paying for it is, in economic parlance, called a **free rider**. In other words, if a participant spots a free rider in this game, they have the ability to punish the free-rider (Fig. 4.10).

[13] This is called a public goods game (with punishment opportunities). While the illustrations in Figs. 4.9 and 4.10 appear to represent a two-player game, the experiment was actually conducted in groups of four. See the end-of-chapter supplement for more information on this game.

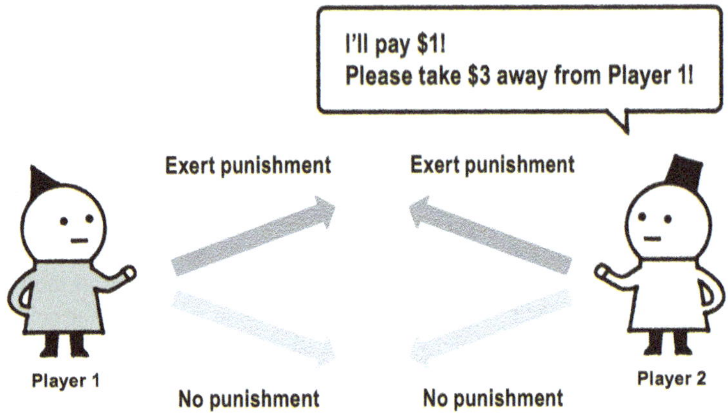


**Fig. 4.10** Altruistic punishment

Hold on, though. The ability to punish is not itself free either! The punisher must pay a cost to exact that punishment. That is, to confiscate money from someone else, the aggrieved must also part with a sum of their own money. And it is crucial to note that any money confiscated in this manner is always forfeited to the researcher. The punisher sees not a cent of it.

### Experimental Results

Turn your attention to Fig. 4.11a. First, the researchers ran the game repeatedly without the opportunity for punishment. When they did this, they found that cooperative behavior became less and less common over time. Having, it seems, figured out that there was no threat of reprimand, the participants ended up just being selfish.[14]

As soon as six without-punishment rounds were complete, however, punishment was introduced, and at this point participants' behavior changed dramatically. Cooperation rates suddenly jumped. And what's more, cooperative behavior steadily increased with each subsequent round. Incidentally, most of the punishments in this phase were issued by participants who were above-average cooperators to those who were poor cooperators—or in other words, free riders.

[14] A decent number of similar experiments have shown that in the absence of punishment, around 40–60% of participants start out cooperating, after which the cooperation rate declines significantly, similar to the results discussed here (Ledyard 1995), and that, eventually, around 70% of participants stop cooperating altogether (Fehr and Schmidt 1999).

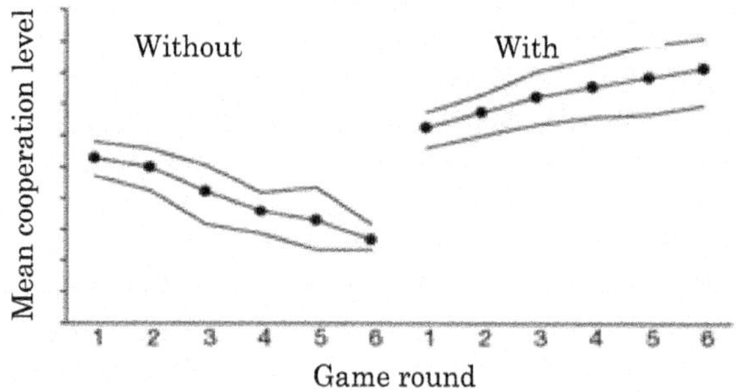


(a)

**Fig. 4.11a** Mean cooperation levels (without-punishment rounds followed by with-punishment rounds). *Source* Fehr and Gächter (2002)

These results tell us that self-sacrificing cooperative behavior, which is costly to the individual, gradually decreases in the absence of punishment. But as soon as punishment is introduced, cooperative behavior increases. Moreover, it increases with each successive round, eventually resulting in a high rate of cooperation.

That's not all. Take another look at the graph. Cooperation increases from the very round in which punishment is introduced. No punishments have even been dealt at this point, yet the mere fear of punishment is enough to prompt cooperative behavior.

Figure 4.11b shows what happened when the order of the without-punishment and with-punishment rounds was reversed. Here too, a very similar pattern emerged.

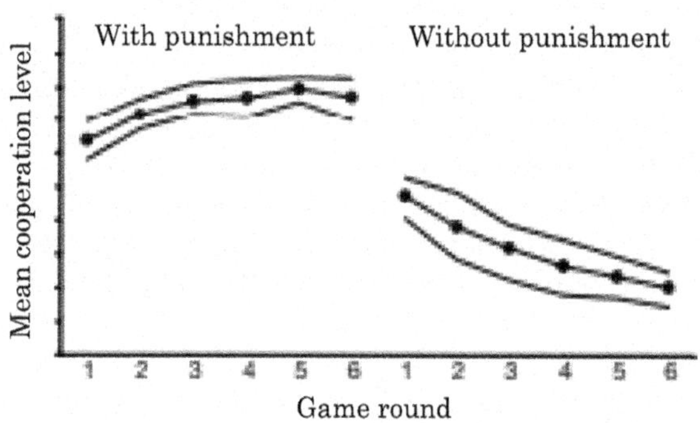


(b)

**Fig. 4.11b** Mean cooperation levels (with-punishment rounds followed by without-punishment rounds). *Source* Fehr and Gächter (2002)

These results tell us that punishment has the power to encourage cooperative behavior. The type of punishment used in the experiment discussed here is called altruistic punishment because, although the individual who chooses to deal out the punishment must do so at a cost, it does encourage socially desirable behavior.

The meaning of the word "altruistic" can be quite nuanced, however, and may at times imply a desire or some sort of motivation to help others. So I would like to set out some terminology for the rest of this book. I will refer to any punishment dealt out by someone who has been harmed in some way as **second-party punishment**. When such second-party punishment constitutes altruistic punishment, I will call it **costly second-party punishment**, with altruistic punishment being referred to as **costly punishment** (Rand et al. 2009). (See the end-of-chapter supplement for a discussion on the usage of the terms "altruistic" or "altruism" and "cooperation").

**No benefit, no matter**

In the game described here, group membership was reshuffled in every round of the game. No two players ever met each other more than once, and no one had any knowledge of any other player's past decisions. So even if you, as a player, were to witness another player being uncooperative in one particular round, you would never come across them again. Reasoning like "If I don't take a tough stance now, they'll just do it again!" doesn't really make sense. Even if someone you punish were to subsequently feel remorse and change their behavior, you would not directly benefit from this.

Plus, even if you were to pay the cost of punishing a free rider for uncooperative behavior, that information would not be passed on to any subsequent partners in the game. No one would look at you and think "This person was willing to pay a cost to punish a bad actor, so I know I can trust them!".

In other words, it's very difficult to see how punishment in this scenario could possibly benefit you at all. The cost of exacting the punishment is the only direct consequence to yourself, and thus an Econ naturally would never do anything so wasteful. Yet over 80% of participants punished someone at least once. And more than one in three dealt out punishments in five or more of the six rounds of the game.

Researchers also found that even when subjects were not informed of whether and how much they had been punished until after the conclusion of all rounds of the game, thus precluding any possibility of benefit to those inflicting the punishment, participants still elected to punish others (Fudenberg and Pathak 2010).[15]

---

[15] Note, however, that even though punishment may encourage cooperative behavior, it does not necessarily follow from this that people punish *in order to* elicit cooperative behavior in the future.

Say, for example, that punishment is imposed to reduce bad behavior in the future. Then the punishment should be harsher when the likelihood of the bad behavior being detected is lower. Experimental evidence, however, shows that the magnitude of punishment is more sensitive to the severity of the offense and extenuating circumstances (mitigating or exacerbating circumstances often affect the moral outrage felt) than to the likelihood of detection (Carlsmith et al. 2002).

In other words, punishment is less about discouraging bad behavior in the future and more about inflicting harm on wrongdoers. This is the just deserts rationale for punishing past misdeeds (Carlsmith et al. 2002). Punishment seems to be backward looking. As evolutionary ethicist Scott M. James puts it: "We do not punish because it deters. We punish because the punished deserve it" (James 2010, p. 73).

It's not that there is no forward-looking aspect to punishment, which would mean, for instance, that it aims to redress unfairness caused by bad behavior (Bone and Raihani 2015), but the existence

Fehr's experimental work also showed that the more selfish the behavior, the more severe the punishment meted out, and that greater transgressions were met with greater punishments. Research has also shown that punishment decreases as the cost of punishing increases (Anderson and Putterman 2006). That is, there is a rational component to costly punishment decisions that reflects the severity of the wrong and the magnitude of the cost.[16]

**Those freaking free riders**

In Fehr's experimental work, participants also reported stronger feelings of anger at those who chose to act selfishly. In a post-experiment questionnaire, participants were asked to rate their anger at free riders on a scale of 0–7, with 7 indicating the most intense anger. Around 50% of people scored their anger a 6 or 7, and around 40% of people gave an anger score of 5. So it is thought that this type of punishment, which does not offer any material benefit, even in the future, is motivated by anger.

This sort of costly second-party punishment, whereby people sacrifice something of value to wield the punishment, has been observed in many subsequent studies as well (Balliet et al. 2011; Gächter et al. 2008; Gürerk et al. 2006; Rockenbach and Milinski 2006; Yamagishi 1986; Herrmann et al. 2008).

We know that retaliation—if you do it to me, I'll get you back!—is widely observed in the animal kingdom as well as in humans (Jensen et al. 2007b; Ohtsuki 2014; Leimgruber et al. 2016). Research also suggests that negative direct reciprocity (getting someone back for being mean to you) develops in humans at an early stage, around the ages of 4 to 8, whereas positive direct reciprocity (being nice specifically to those who are nice to us) does not develop until a little later (Chernyak et al. 2019).

Retaliatory behavior—getting someone back if they harm you—may certainly seem costly on an immediate time horizon, but it may help us protect ourselves over the longer term by serving to reduce attacks from others in the future (Kobayashi 2021). In any case, leaving all the sophisticated reasoning aside, it is, perhaps unfortunately, pretty easy to relate to the desire to pay someone back if they've done something nasty to us.

---

of this backward-looking aspect of punishment is something I think we would do well to keep in mind, particularly in the context of our modern online culture.

[16] Some researchers have reported that above-average cooperators, instead of uncooperative agents, are often punished (Hermann et al. 2008; Gächter et al. 2010). This somewhat odd behavior of punishing highly cooperative people is called **antisocial punishment**. The likelihood of this happening is highly culturally dependent (Bruhin et al. 2020). It is apparently more likely to occur in countries with weak social norms around cooperation, such as those where behavior like tax evasion and public transport fare evasion is more readily tolerated. Antisocial punishment can discourage cooperation (Hauser et al. 2014), and it is sometimes explained as an attempt at retribution against people who are likely to mete out punishment, but its motivations are not necessarily all that well understood. Researchers have also reported that the brain's response to antisocial punishment may differ to its response to the punishment of defectors (Gerfo et al. 2019).

## 4.3 Sweet, Sweet Punishment

### Empathy, an ancient system

The act of punishing generally carries quite a high cost. It puts the punisher at risk of incurring injury or enmity, for instance. In the ultimatum game, this cost corresponds to sacrificing your own share of the pot. Why is this behavior so widely observed when it comes at a high cost and offers no obvious material benefit to the punisher. To get at the brain mechanisms that underpin costly second-party punishment, we'll first take a look at the mechanisms underlying human empathy.

### I feel your pain

**Empathy** is what we call the feeling we experience when we see someone else cut their hand and think "Ow, that looks painful!" or when we ourselves start to feel nervous at the sight of someone else standing precariously on the edge of a cliff.[17] While he didn't use the term "empathy" (English had no such word at the time), Adam Smith, often called the father of economics, explained this feeling masterfully in his book *The Theory of Moral Sentiments*: "By the imagination we place ourselves in his situation, we conceive ourselves enduring all the same torments, we enter as it were into his body, and become in some measure the same person with him, and thence form some idea of his sensations, and even feel something which, though weaker in degree, is not altogether unlike them." (The term "empathy" later came into English, although not exactly with its current meaning, as a translation of the German word "Einfühlung", meaning "feeling-in" or "feeling into". The word Smith used was "sympathy".)

To explore what it means to "become the same person", as Smith puts it, let's think about an experiment. Imagine, for instance, that you are first given a painful electric shock. And then the same electric shock is administered to your partner sitting across from you. Research indicates that upon witnessing this second shock, parts of your brain that are activated when you experience pain—including the **anterior insula** and **anterior cingulate cortex** (ACC)—also light up when you see your partner in pain and despite you not receiving a direct pain stimulus yourself (Singer et al. 2004). Simply witnessing someone else experience pain is enough to involve the parts of your brain that respond when you experience that sensation first-hand.

Work in a variety of other scenarios has also demonstrated commonalities in the brain's response to our own experience of pain and its response to seeing, or otherwise "observing", others experiencing physical pain—this of course holds when we directly witness something painful, like someone being pricked with a pin, struck, or exposed to heat, and it also holds when we view facial expressions contorted by

---

[17] It has become common in recent years to think about empathy in two broad categories: **emotional empathy**, whereby we have a visceral sense of what others are feeling, and **cognitive empathy**, whereby we understand, intellectually, what others may be feeling. Here I am of course discussing emotional empathy.

Emotional empathy is also observed in some animals, as discussed, but cognitive empathy can be regarded as perhaps unique to humans (Hasegawa 2016).

pain or read someone's account of experiencing pain (Lamm et al. 2011; Bloom 2016; Fallon et al. 2020). In sum, this research tells us that other people's pain activates regions of the brain responsible for our own pain response (see the end-of-chapter supplement for more information).

Children also experience this empathetic response to the pain of others (Decety & Michalska 2010), and research has shown that infants as young as 1 or 2 years old, and even newborns, become distressed (which may include whimpering, whining, or crying) when they see someone else suffering (Sagi and Hoffmann 1976, Dondi et al. 1999, Bloom 2016).

These findings led De Waal (2013, p. 53) to conclude that "Insects lack empathy, whereas our brains are built to connect with others and experience their pain and pleasure."[18]

**Empathy in animals**

Empathic reactions have also been reported in the animal kingdom (for a review, see de Waal 2019). Researchers have found, for example, that rats and mice may have an empathetic pain response—as determined by body movements typical of a pain response, brain activation, etc.—to seeing their cagemates experience pain (Langford et al. 2006; Carrillo et al. 2019). Others have found that certain rodents may not only experience empathy but also display consolation behavior toward distressed cagemates (Burkett et al. 2016), and that rats will put priority on helping a cagemate over gaining access to chocolate, which they love (Bartal et al. 2011).

Some of the earliest of these sorts of studies of empathy in animals showed that rats (Church 1959), monkeys (Masserman et al. 1964), and even pigeons (Watanabe & Ono 1986) stop taking food when this act results in a conspecific (a member of the same species) in a cage next to them receiving an electric shock. The monkeys even went up to 12 days without eating. More recent work has demonstrated that rats, like humans, are reluctant to hurt others (Hernandez-Lallement et al. 2020).

Charles Darwin, the renowned British naturalist and father of evolutionary theory, was referring to what we now call empathy when he wrote as follows in *The Descent of Man, and Selection in Relation to Sex*: "Both the appreciation and the bestowal of praise and blame rest on sympathy; and this emotion, as we have seen, is one of the most important elements of the social instincts." And, similarly, in the same book: "Nevertheless the first foundation or origin of the moral sense lies in the social instincts, including sympathy; and these instincts no doubt were primarily gained, as

---

[18] Recall the work, discussed in Sect. 3.3, showing that isolation and ostracism are painful. That research shows that regions of the brain implicated in your own physical pain are also involved in your response to social pain. Putting all of this together, we can say that your own physical pain, your own social pain, and the physical pain of others all look very similar to the brain.

The following question then naturally arises: What about others' social pain? Things are actually a little different here. Studies have shown that regions of the brain collectively called the **mentalizing network** (MN), which includes brain regions involved in cognitive empathy, are predominantly activated in this case (Masten et al. 2010; Masten et al. 2011). Researchers have argued, however, that the social exclusion of our friends and others we are close to may in fact activate the anterior insula and anterior cingulate cortex, which are involved in emotional empathy (Meyer et al. 2013).

in the case of the lower animals, through natural selection." (Darwin 1871, p. 393 & p. 394). Over the course of more than a century since, evidence has continued to mount that empathy developed in animals even before our ancestors became recognizably human.

If you have children, you will no doubt understand that the power of empathy is what constitutes one of the true joys of being a parent. To be entranced by the beaming face of a child and experience anew the utter glee of seemingly trivial events like eating ice cream, jumping into the sea, or creating a towering stack of building blocks is to experience the true power of empathy. And I'm sure this sort of feeling will ring a bell with anyone able to recall the joy of someone close to them.

**Empathy in evolution**

Prominent psychologist Paul Bloom writes: "I favor Batson's own analysis that empathy's power lies in its capacity to make the experience of others observable and salient, therefore harder to ignore." (Bloom 2016, p. 95). In other words, by making us quiver at the fear someone else may be feeling or having us think "Ouch, that must hurt!" when someone else is in pain, empathy has the ability to let us know that someone is suffering. It has thus been argued that the roots of empathy's power in this sense lie in the mammalian mother–child relationship since it is crucial that mammalian mothers are able to quickly detect distress in their offspring.

By allowing us to make the suffering of others our own, empathy can also pave the way to compassion, the desire to alleviate the suffering of others. Empathy, in that sense, is considered a key part of the foundations underpinning human morality and prosocial behavior.

There is at least one other important aspect to empathy's power. When someone else shows fear or distress, this acts as a signal to you that something is afoot. The ability to automatically empathize with your fellow humans when they are in distress also gives you the ability to quickly sense danger and thus protect yourself. Indeed, it is quite possible that this is actually the origin of empathy in an evolutionary sense.

De Waal (2014, p. 13) thinks that "cooperation may be motivated by empathy, a characteristic of all mammals, from rodents to primates. We identify with others in need, pain or distress." And Scott M. James (2010, p. 127) speculates that "Perhaps distress-sensitivity is an ancient system."[19]

---

[19] Although empathy is generally seen as a very good thing, this is not always the case. Empathizing with someone who is suffering, for example, may cause us such distress ourselves that we end up choosing to avoid the poor soul rather than help them.

We also tend to feel empathy much more readily toward people close to us (in-group members) and much less toward those who are remote from us (out-group members). So we know that empathy can explain our tendency to think only of those who are close to us as important (Bloom likens it to a "spotlight focusing on certain people in the here and now"). With work in this area gaining more and more attention in recent years, researchers have also pointed out the negative aspects of empathy. The interested reader is encouraged to consult Bloom's *Against Empathy: The Case for Rational Compassion* (Bloom 2016).

## Devil in Our Heart

Now, it's all well and good to say that we are able to feel others' pain as if it were our own, but we also know that this feeling of empathy can spontaneously evaporate in certain circumstances (Singer et al. 2006). The experiment we'll look at in this case involved a total of 32 adults, 16 men and 16 women.

The participants played several rounds of a game in groups of three, but each subject was (unbeknownst to them) always set up with two confederates (professional actors collaborating in the study). One confederate was the fair player, who divided the money generously, while the other played the part of the unfair player, who was tight with the money. Once the game rounds were complete, the researchers ran the electric shock experiment described earlier and measured the subject's brain response to seeing the other players experience pain.[20]

As expected, participants' responses to the fair players' suffering, as with the aforementioned experiment, were suggestive of empathy. That is, they saw the fair player suffering and thought something like "Ooh, that must hurt!".

Participants' responses to the unfair players, however, were completely different. Take a look at Figs. 4.12 and 4.13, the former showing the response in women, and the latter the response in men. The areas circled in the brain images (a and b) show the brain regions activated, and the graphs (c and d) show the left and right brain responses.

Starting with the women's responses in Fig. 4.12, the researchers observed a reduction, albeit a small one, in activity in the brain's pain network on both the left and right in response to seeing unfair players in pain.[21] The picture was a little different in men, however, as Fig. 4.13 shows. The empathetic response in men was so diminished as to have seemingly disappeared.[22]

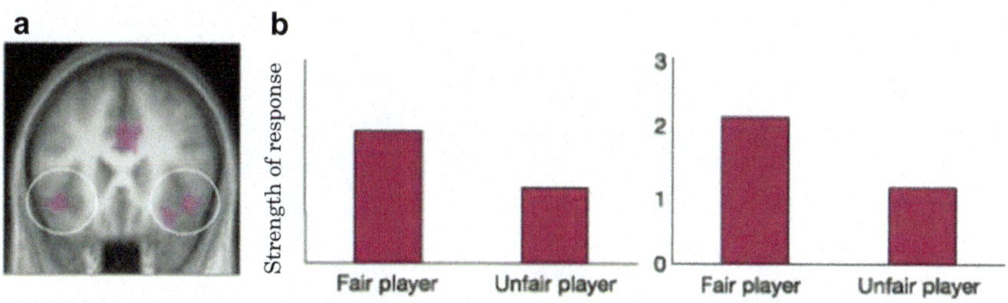


**Fig. 4.12** Brain response in women to the sight of pain in fair and unfair players. *Source* Singer et al. (2006)

[20] The game player here was a sequential prisoner's dilemma game (PDG). For details on the game's structure, see the end-of-chapter supplement.

[21] The drop in response was not statistically significant, however, meaning that the change was not big enough for us to rule out the possibility of it simply happening by chance.

[22] Researchers have also reported that people show greater empathy for the pain of fans of the same sports team and people of the same race as themselves, whereas there is not a whole lot of empathy for fans of rival teams or for people of a different race (Xu et al. 2009, Hein et al. 2010). We humans, unfortunately, do perhaps have a tendency to focus our empathy solely on those in our peer group.

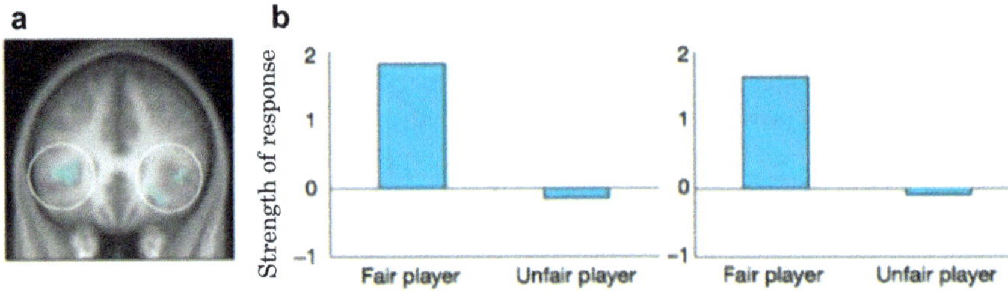


**Fig. 4.13** Brain response in men to the sight of pain in fair and unfair players. *Source* Singer et al. (2006)

## The Joy of Punishment—That's What You Get!

That's not all. It may come as a surprise that in the men who participated, the brain's reward system (striatum) was also activated!

You may recall the reward system from earlier in the discussion—the part of the brain responsible for joy and pleasure. Look at Fig. 4.14. This graphs the strength of the brain's reward system response to the sight of an unfair player in pain. The left bar corresponds to the response in women, and the right bar to the response in men.

The graph indicates high activation in the mens' brain reward system when observing an unfair player receiving painful stimulation. Seeing a "bad guy" suffer feels great[23]! This reward system activation was also stronger in men who expressed a stronger desire for revenge. That is, such individuals experienced greater joy and pleasure when seeing a disdainful person suffer (In the image, "N.acc." refers to the nucleus accumbens, part of the ventral striatum, an important part of the reward system).

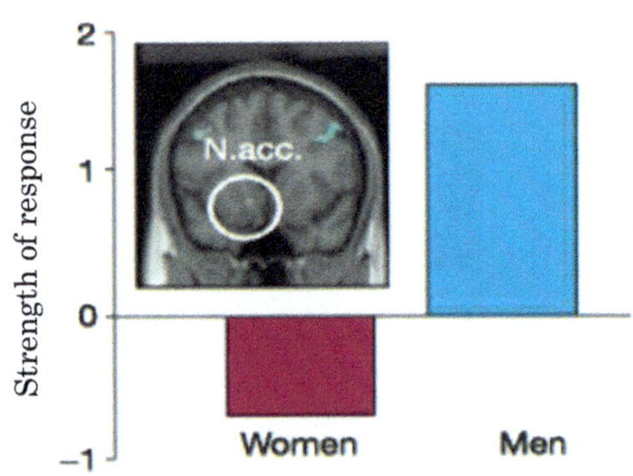


**Fig. 4.14** Brain response indicating joy/pleasure at seeing an unfair player in pain. *Source* Singer et al. (2006)

[23] Other research indicates that the brain's reward system (striatum) is activated in response to seeing a fan of a rival sports team experience pain (Hein et al. 2010), and, as discussed in a footnote in Sect. 4.1, that the reward system is activated in response to misfortune suffered by an enviable person (Takahashi et al. 2009).

An analysis of facial muscles also showed that people smile when witnessing an enviable person beset by misfortune (Cikara and Fiske 2012). It has also been pointed out, though, that people who smiled did self-report that they felt a little bit bad about such misfortunes suffered by others, which is quite inconsistent with what the analysis of facial expressions indicated.

We know that humans are very reluctant to intentionally cause pain to others (Cushman et al. 2012). Work following the Cyberball experiment discussed in Sect. 3.3 showed that individuals who complied with instructions to ostracize others without reason also experienced pain (Legate et al. 2013). So at any rate, we know that empathy gives us the ability to make others' pain our own. In light of this, the finding that people are able to derive joy and pleasure from witnessing the pain of others is pretty shocking.

The application of neuroscience to economics is called **neuroeconomics**. These days, though, it may make more sense to talk about **social neuroscience** and the **social brain**. This research is what taught me the importance of neuroeconomics and social neuroscience. Perhaps it might have been possible to conduct a questionnaire asking people how they felt at seeing someone receive a painful electric shock. But I surely doubt that people would give us responses like "That's just what they deserve!" or "I was quite pleased to see them suffer!" It was seeing the brain responses first-hand that taught us about the pleasurable response to punishment. This research is what made me realize that the collaboration with neuroscience is indispensable to the humanities and social sciences going forward.

**A brief aside: Do boys worship heroes?**

I once heard an argument that the ability of the male brain to find joy in the punishment of bad guys is reflected in differing gender preferences for children's TV programs like action hero or superhero shows. Come to think of it, my daughter has little taste for them.[24]

No doubt such shows appeal to those who rejoice, and perhaps even cheer, at seeing bad guys receive a pummeling. But what if you naturally feel empathy for the bad guys or monsters or mutants or whatever? What if you think it's a bit mean to rain death and destruction upon some outsider just for showing up at Earth uninvited? Well then such shows might not be all that appealing to you. I feel there is definitely a chance that the brain characteristics we've been discussing are reflected in people's TV preferences. (even so, I may have watched my fair share of superhero shows as a little girl with my big brother; alongside, I also played dolls and enjoyed playing house with my sister too).

That said, I also don't think the upshot of all this research is that women are simply kind and lovely! Although the research on aggression provides a lot of data about aggression in men, it's not that women are bereft of aggression. Women are quite capable of being aggressive when competing for valuable resources (Kenrick 2011). Research in this area also highlights the possibility that male and female aggression differs in its fundamental nature. Men may prefer a direct approach to aggression manifested in the form of physical violence and overtly foul language, whereas

[24] Actually, our sons do not enjoy superhero shows either. It is absolutely crucial to acknowledge the existence of individual differences and diversity. In addition, even within one person, there exist multiple aspects. Unfortunately, this book does not touch upon topics of such diversity due to space limitations, but please note that this does not mean that these perspectives are unimportant.

women perhaps prefer indirect channels of aggression that include ostracism, innuendo, and gossip (Archer 2004; Hess and Hagen 2006; Fields 2016). (Research has also shown, however, that while people in positions of social power may express anger directly, those in relatively powerless positions are more inclined to express anger indirectly for fear of reprisal (Petkanopoulou et al. 2019). So it may also be possible that the way aggression is expressed is more a reflection of social position and status than of male–female differences.)

I also think that women might react a little differently in experiments that employ punishments of the type that they are more acutely averse to—think mental torture and anguish—as opposed to the, in some sense, pedestrian approach of administering physical pain via electric shocks and so forth. Women are not to be trifled with, as they say. Hey if anyone wants to collaborate, come on!?[25]

**Revenge is sweet**

So, how does the human brain react to costly second-party punishment?

As you might predict from experiments indicating that some people relish the sight of bad guys suffering, research has also shown that the brain's reward system (striatum), the part responsible for joy and pleasure, lights up in response to the punishment of people for deliberate violations (De Quervain et al. 2004) (Fig. 4.15).[26]

**Fig. 4.15** Second-party punishment and the joy of punishment

What's more, this brain activation was only observed when the punishment actually harmed the wrongdoer, or in other words, when it was an effective punishment. The part of the brain implicated in pleasure and joy did not respond when the punishment was merely symbolic and thus did no real harm—verbal criticism, for instance.

---

[25] While we've discussed the distinction between males and females as a premise in this book, let me add as a reminder that it is inherently very important to recognize that, in fact, men and women, whether in the biological sense or not, cannot be discussed solely in terms of such a simple dichotomy.

[26] The game used in this case is called the trust game. See the end-of-chapter supplement for details on the game's structure.

This may indeed reflect a strong desire in humans to give bad actors a real taste of suffering.

The researchers also found that the greater the reward system activation in a subject, the bigger the punishment they dealt out. Remember that this reward system also lights up when we eat something sweet. So revenge really is sweet, as they say (Knutson 2004).[27]

You may be somewhat relieved to learn, however, that when the other person's bad behavior was unintentional, the part of the brain that registers joy and pleasure in punishment was not activated. That is, people don't seem to be thirsty to see others suffer for inadvertent transgressions.[28]

## 4.4 Third Party Punishment

### You Can't do That!

So far, I've talked about how humans dislike intentional unfairness and show a strong desire to punish those who engage in such behavior. The human desire for punishment does not end here, however. We seem to have a strong desire to punish wrongdoers even when the situation apparently has nothing to do with us, and this of course goes beyond the fairly instinctive desire to pay people back when they've harmed us. Let's take a brief look at this interesting aspect of human behavior.

Suppose that in the dictator game, Player 1 divides the pot as follows: ($9.99, $0.01). Now suppose that a third-party bystander, whom we'll simply call Player 3, has the ability to pay a sum out of her own pocket to fine Player 1 three times that amount. That is, if Player 3 pays $1, she can take $3 away from Player 1. If she pays $3, she can take $9 away (Fig. 4.16).

[27] Immediately after dealing out a punishment, however, I think people generally also feel a somewhat difficult-to-describe aftertaste of sorts, even if they find the initial act of punishing to be sweet. Indeed, research has also shown that people who issued a punishment experienced negative feelings afterwards. For details, see the end-of-chapter supplement.

[28] The presence of intent also seems to affect the way we feel pain, for instance (Gray & Wegner 2008). Researchers found that the pain of repeatedly administered electric shocks gradually diminished when shocked subjects believed there was no intent behind the decision to expose them to the punishment, but that the pain persisted when they believed there to be intent behind the punishment. To me, this finding that intentional mistreatment hurts a lot more makes a lot of intuitive sense.

**Fig. 4.16** Costly third-party punishment

Any amount taken away from Player 1 is still confiscated, however. Player 3 receives none of the proceeds of the fine, and so Player 3 does not herself benefit by punishing Player 1. And the way money is divided between Player 1 and Player 2 has nothing to do with and no bearing on Player 3 either.

It is costly to issue a fine and thereby punish Player 1, so an Econ would, of course, see it as pointless and never actually fine Player 1. Yet in an experiment of this type, around 60% of people in Player 3's position paid money out of their own pocket to punish Player 1, despite Player 1's action having nothing to do with Player 3 (Fehr & Fischbacher 2004).

**Costly third-party punishment**

When a third party who was merely watching on and not directly harmed metes out punishment like this, it is called **third-party punishment**. When the third party pays a price to deal out an altruistic punishment to a selfish person, I will call this **costly third-party punishment**.[29] In the experiment just described, Player 3, who was not harmed directly and was instead merely a third-party bystander, went to the trouble of paying a cost in order to punish Player 1 for being greedy. Player 3 here is of course engaging in costly third-party punishment.

[29] It must be noted that the term "third-party punishment" is often used by itself to refer to what I am calling costly third-party punishment here; that is, altruistic punishment whereby one party sacrifices something of value to punish selfishness. The usage of these terms varies across disciplines, it seems, but in this book I use the term "third-party punishment" irrespective of whether the wielder of punishment had to pay a cost (this mirrors the usage of "second-party punishment" earlier in the discussion). When the wielder of punishment pays a cost, I make the distinction clear by using the term "costly third-party punishment".

Costly third-party punishment has been studied extensively in the lab (Fehr and Fischbacher 2004; Ohtsubo et al. 2010; Jordan et al. 2016a; Yamagishi et al. 2017; Kamei 2018; Kamei 2020), but it has not remained confined to this setting. It has also been observed through experiments in small-scale societies that earn a living by traditional means[30] (Henrich et al. 2006; Bernhard et al. 2006; Henrich et al. 2010), among members of the Swiss Army (Goette et al. 2006), and in field experiments designed to study more realistic, natural conditions (Balafoutas and Nikiforakis 2012; Balafoutas et al. 2014a; Artavia et al. 2017).[31] Research has also shown that the brain's reward system (striatum) is activated by third-party punishment too (Strobel et al. 2011; Stallen et al. 2018) (Fig. 4.17).

**Fig. 4.17** Third-party punishment and the joy of punishment

Children also display similar tendencies. Researchers have reported that children around the age of 6 (McAuliffe et al. 2015a; Jordan et al. 2014; McAuliffe and Dunham 2021) and even children as young as 3 to 4 years old (Yudkin et al. 2020) engage in costly third-party punishment. They will, for instance, punish a friend for doing something bad, even if they are not themselves directly affected, by preventing the use of a toy even at the cost of giving up the use of that toy themselves. More recent research suggests that even preverbal infants as young as 8 months may use third-party punishment (Kanakogi et al. 2022).[32]

[30] Note, however, that experiments in small-scale societies indicated that third-party punishment becomes more likely as the size of the society increases, and that third-party punishment is less likely in small societies (Marlowe et al. 2008; Marlowe 2009).

[31] It is crucial to note, however, that the severity of costly third-party punishment observed in field experiments is significantly lower than in laboratory experiments. For details, see the end-of-chapter supplement.

[32] Third-party punishment has not been observed in experiments with chimpanzees (Riedl et al. 2012) and has not been reported in non-human animals (Jensen & Tomasello 2019).

This tendency to step in as a third party and punish other people for behavior that does not directly affect you or otherwise have anything to do with you is not entirely unrelated, it seems, to the problem of online bullying and abuse that has raised its head in recent years[33].

## A Third-Party Perspective—I'm Not Standing by to Watch you Cheat and Lie

Engaging in third-party punishment requires the ability to evaluate the actions of others with whom you have nothing to do directly. Research in recent years has found that this ability can be traced back not only to the age of 3 (Vanish et al. 2010) but also as far back as infancy (Hamlin et al. 2007).[34]

In this study, infants were shown two scenes, one in which a character helped a climber ascend a hill and one in which a character hindered the climber from achieving that goal. The infants then picked either the helper or the hinderer. Both 6-month-old infants and 10-month-old infants chose the helpful character. The infants, it seemed, shunned individuals who were mean to their friend. Subsequent research also showed a similar tendency in very young children of only 3 months of age (Hamlin et al. 2010).

You might think that these babies simply preferred people who were kind to others, but this may not be the whole picture. Research indicates that infants as young as 8 months old may prefer characters who act positively toward prosocial individuals and characters who act negatively toward antisocial individuals, suggesting complex reasoning (Hamlin et al. 2011). Somewhat more recently, researchers have reported that infants as young as 6 months old may show similar tendencies (Kanakogi et al. 2017). In the words of Yasuhiro Kanakogi, who conducted this research, "In human society, selflessly protecting the powerless is considered an act of heroic justice...

---

Interestingly, however, even in the non-human animal world, high-ranking individuals may stop conflict between third parties, a behavior known as policing. Policing has been reported in a wide range of species, including bonobos, gorillas, macaques, and baboons (Von Rohr et al. 2012, Beisner and McCowan 2013, Tomasello 2019).

[33] Some have speculated that the motivation behind third-party punishment is anger (Jordan et al. 2016b), but with the nature of that anger being a little different from that of Player 2, who is directly affected by unfair actions. Others have reported that disgust and sadness are strongly associated with both second-party punishment and third-party punishment, that second-party punishment is more strongly associated with anger, and that third-party punishment is more strongly associated with a sense of moral outrage (Hartsough et al. 2020). Research has also revealed that while there are commonalities in the brain reactions associated with second-party punishment and third-party punishment, differences are also present: third-party punishment preferentially engages social cognition regions of the brain, for instance (Bellucci et al. 2020).

Researchers have also reported that rewarding punishers—i.e., paying people to engage in moralistic punishment—cut their willingness to do so in half (Rai 2022). This sort of work continues to probe the emotions and mechanisms underlying punishment behaviors.

[34] However, this study has faced criticism regarding its methodology (Scarf et al. 2012) and reports of failed replication attempts (Schlingloff et al. 2020). Particularly noteworthy is a study published during the proofreading of this book, which reported that a large-scale experiment, including the original author, failed to replicate the findings (Lucca et al. 2025). Ongoing attention to advancements in this field, along with related research, is of utmost importance.

[and the findings suggest that] recognizing heroism is perhaps an innate ability" (from a press release published on Kyoto University's website).

A recent meta-analysis has confirmed this finding that infants prefer prosocial agents, or simply "nice people" (Margoni and Surian 2018). The analysis looked at data on infants aged 4–32 months, and the investigators did not find that the preference became stronger as age increased. This may mean that this is not a learned preference, or in other words, not something that children acquire as they grow up.

Research has also shown that from the age of 19 months, toddlers tend to punish puppets who were mean to a friend, with the punishment being the act of taking a treat away, and it is thought this may represent the beginnings of our desire to punish (Hamlin et al. 2011).[35]

This ability to evaluate third-party interactions has also been observed in capuchin monkeys (Andersen et al. 2013a; Andersen et al. 2013b), marmosets (Kawai et al. 2014), and even dogs (Chijiiwa et al., Anderson et al. 2017).[36]

## 4.5 Double-Edged Sword Wielded by an Invisible Hand

**Pre-agriculture life led by the invisible hand**

We humans can empathize with strangers. We have the capacity to make someone else's pain our own. But if we decide, for whatever reason, that someone is a bad guy, not only does our empathy suddenly evaporate, but we also take joy and pleasure in seeing the miscreant suffer. Matteo Motterlini, known for his contributions to the philosophy of science and the history of ideas, refers to this, in a sense, sinister and formidable mechanism within our brains as the **invisible hand** (Motterlini 2008).

---

[35] Children were previously thought of as passive norm followers in that they learn social norms by following cues from adults, yet they are now considered to be active norm enforcers. A slew of research has also shown that children quickly discover norms by observing the behavior of those around them, and that, even when they are a third party, children actively reprimand norm violators (Rakoczy et al. 2008; Rakoczy et al. 2009; Schmidt et al. 2012; Schmidt and Tomasello 2012; Schmidt et al. 2016; Henrich 2017).

In such studies, the researchers found that children as young as 3 years old quickly figure out prevailing norms by observing others around them, believe that those norms are something everyone must abide by, immediately express anger at norm violations and violators, offering up protests like "No, that's not how you're supposed to do it!", and show a desire to teach the correct norms.

Even more interesting is that these behaviors don't necessarily seem to require direct modeling or adult guidance. Even in the absence of admonishments or warnings from an adult, or instructions like "If someone does something the wrong way, you have to tell them off", children will spontaneously reprimand norm violators, even when they are only a third party and even when interacting with rule-breakers they've only just met for the first time (see Tomasello 2019 for the details).

[36] As mentioned in footnote 56, behavior such as rewarding the good and punishing the bad based on such third-party evaluations has not been observed in non-human species. Interestingly, however, researchers have reported an absence of third-party evaluations in Japanese monkeys, which are generally considered to be high in aggression and low in tolerance (Kawai et al. 2019).

Some seven million years have passed since our species diverged from its common ancestor with chimpanzees in Africa to form the human lineage, and perhaps some 200,000 years have passed since that species became *Homo sapiens*, what we are today.[37] Agriculture has only been around for about 10,000 years. So if we compress the entire human lineage into a single year on the calendar, *Homo sapiens* only appear in late December, and agriculture only pops up on December 31. Our lineage has spent the bulk of its history living a pre-agricultural existence.

Before the advent of agriculture, *Homo sapiens* are thought to have lived in relatively small groups of around 150 people, obtaining food through hunting and foraging. *Homo sapiens* run slowly and are not armed with strong fangs. Even today, if you were to somehow find yourself stranded alone on the African savannah, you would probably not survive all that long.

Prominent anthropologist Donald E. Brown, who has studied human cultures around the world, calls the characteristics that are common across human societies "human universals" (Brown 1991).[38] One of the human universals identified by Brown is that we are "not solitary dwellers"—we spend most of our lives living in groups. Forming groups was an important survival strategy for humans.

The ability to empathize is crucial to creating and maintaining such groups. Empathy is the ability to feel what others are feeling, and without this, maintaining group cohesion would be pretty difficult.

Yet having nothing but empathy would also be problematic. People who break group rules to their own benefit can be called free riders. Punishing is the act of causing someone pain of some sort. So if you were only able to empathize with the pain of others, you would find it fairly difficult to punish any free riders in your midst. And without punishment, free riders have the run of the place to do as they please. Before long, the group would no doubt fall apart. There is a major dilemma here: while empathy is crucial to group life, empathy by itself would effectively preclude us from punishing free riders.

This is where this invisible hand comes into play. While empathy helps to maintain group cohesion, the invisible hand lets us set aside empathy for selfish free riders and, moreover, to take joy and pleasure in the suffering of such "cheaters", enabling us to quite happily punish them. And even people in a third-party position who have no direct interest in such punishment (nothing to gain or lose directly from it) are able to derive joy and pleasure from it. Whereas we would normally feel others' suffering as our own pain, the invisible hand, as described by Motterlini, gives us the ability to punish free riders.

---

[37] You may have learned that the earliest species on the human branch of the evolutionary tree (these days referred to as hominins) is *Australopithecus*, but some now argue that this spot should go to *Sahelanthropus tchadensis*, the fossilized remains of which were discovered in Chad in 2001. While it is proposed that this is the earliest known species to have appeared after ancestral humans split from chimpanzee forebears, this claim has been contested, and there is a good chance of things changing as our understanding progresses. The *Sahelanthropus tchadensis* fossil found in Chad was nicknamed Toumai, meaning "hope of life" in the local language (Roberts 2011).

[38] The list of human universals is long, but the famous psychologist Steven Pinker offers a concise summary in *The Blank Slate* (Pinker 2002).

The term "invisible hand" is first attributable to Adam Smith. In free markets, says Smith, individuals acting purely in their own self-interest still end up meeting the needs of society as a whole through their actions, as if guided by an invisible hand. Motterlini pointed out that even though we humans might not consciously punish free riders *for the good of the group*, the mere fact that we take joy in the suffering of cheaters and are thus motivated to punish them means that, ultimately, the act of punishing free riders ends up being for the good of the group, as though we were guided by an invisible hand. Everything, in this sense, is a survival strategy developed by our brains over the course of evolution.[39]

**Deviating from the norm**

Recall that characteristics common across human societies are known as human universals. The fact that we punish people who do wrong is so prevalent across human societies as to be listed among such human universals. That is, it really is present all over human societies and observed in all sorts of real-world situations far beyond the lab. Anthropologist Christopher Boehm, for instance, who has studied traditional hunter-gatherer tribes around the world, makes the following point: "when it comes to the really serious political dominators I just mentioned, and also a few other seriously deviant types, a firm and sometimes ultimate line is drawn in the sand; he who crosses this line must be prepared to sacrifice his genetic future" (Boehm 2012, p. 61). Nobel laureate Elinor Ostrom (incidentally, the first woman to win the Nobel Memorial Prize in Economic Sciences) points out that many of the world's common-pool resources have been preserved by sanctions voluntarily agreed upon and imposed by members of the respective communities (Ostrom 1990).

As you will no doubt soon appreciate, however, this invisible hand wields a double-edged sword. What would the world look like if we had an empathetic blind spot that let us take nothing but pleasure in the suffering of anyone labeled a "bad guy"? As one student said to me, "Now I get why there are always wars going on somewhere in the world."

**Summary**

About the responder (Player 2).

- Both humans and animals loathe unfairness
- Humans dislike intentional unfairness in particular
- Humans dislike being looked down upon, seen as inferior
- Can't stand unfair proposals
- Willing to sacrifice something of value to punish unfair opponents
- Empathizes with the pain and suffering of others, but the suffering of bad actors provides joy and pleasure

---

[39] There is still much to think about, however. Punishment may be useful for maintaining group order, but why do people engage in punishment when the punishment itself does not benefit them as individuals? What does an individual ultimately gain by punishing others? For more on the problems of punishment behaviors, see the end-of-chapter supplement.

About the third-party bystander (Player 3).

- Willing to sacrifice something of value to punish a bad actor even when she has no direct interest in doing so
- The suffering of bad actors provides joy and pleasure

To summarize the discussion so far, the responder (Player 2) seems to feel intense disgust and anger at being intentionally treated unfairly and offered only a small share of the pot, and thus shows an apparently strong desire to punish Player 1 as a bad actor since the suffering of those who wrong us is pleasurable. This seems to be one of the motivations for rejecting proposals in the ultimatum game. And this strong desire to punish bad actors does not itself necessarily have anything to do with the punisher having a direct interest in (having something to lose or gain from) the interaction.

## Chapter 4 Supplement

### A brief look at schadenfreude

In Sect. 4.1, we discussed the human dislike for inequity. But I also mentioned that a disdain for inequity could also lead to us rejoicing in the misfortune of those who are better off than ourselves.

Indeed, researchers have reported that observing the misfortune of someone in a superior position to us activates our brain's reward system (striatum); that is, we are happy when people who are better off than us suffer misfortune (Takahashi et al. 2009). You've probably heard the term "schadenfreude", imported from German, used to describe this feeling of joy at the misfortune of others (the word is a combination of the German nouns *Schaden*, meaning "damage" or "harm", and *Freude*, meaning "joy"). Research indicates that people will punish someone in an advantageous position relative to themselves even when that person has done nothing (Deutchman et al. 2021). Perhaps our virtuous dislike for inequity and injustice—what we might think of as a beautiful part of our nature—is inseparable from the competitive spirit that makes losing intolerable and the jealousy we feel at others who are better off than we are.

Some interesting experimental work in mice even indicates that animals might experience schadenfreude (Watanabe 2020). The results of this work suggest that low-status mice may enjoy seeing individuals of higher social status feel pain. An intriguing part of the findings of this study is that the phenomenon was very common in mice observing individuals of only a slightly higher status than their own. Perhaps it is hard for us to immediately put ourselves in the shoes of individuals from vastly different social classes, even when they have suffered a failure. And perhaps this has something to do with the idea that if you want to move up in the world, a slight mistake or misstep by someone just above us on the social ladder may, for better or worse, present you with a golden opportunity.

### Motives for and diversity in second-party punishment

In Sect. 4.1, I discussed the possibility that one motivation we humans have for punishing others is to repudiate the sense of having been seen as inferior. There seems to be more than one motivation for rejecting proposals, however. Researchers have reported, for instance, that there are at least two types of people who punish others by rejecting proposals in the ultimatum game (Yamagishi et al. 2017).

In that study, the researchers identified two types of costly second-party punishers in the ultimatum game: altruistic punishers and spiteful punishers. They then showed there to be morphological differences in the brain between the two types. Specifically, they found that the size of a particular region of the brain differed.

An altruistic punisher is someone who exerts costly punishment not only when wronged herself but also when in a third-party position with no direct interest in the interaction. The researchers found that when such people took on the role of the proposer, they acted altruistically and considered others' interests.

A spiteful punisher, on the other hand, is someone who, when wronged, takes revenge (punishes) even if it costs something, but who does not exert costly punishment when in a third-party position. When in the role of the proposer, such people consistently acted selfishly in an attempt to increase their own share of the pot.

For space considerations, I do not discuss this sort of diversity at all in this book, choosing instead to focus on the most typical behavioral patterns revealed by each experiment in question. No matter what the experiment is, however, it is virtually never the case that you find anything approaching 100% of subjects choosing the same behavior. I think this sort of diversity is an inevitable part of our evolution. It's important to keep in mind that there are many different types of people in the world and that they display a diverse range of behaviors in any given situation.

### Some usage notes on the term "altruism"

As explained in Sect. 4.2, **altruistic punishment** is the act of punishing someone for bad behavior even when it costs you something to do so.

In behavioral ecology, **altruistic behavior** is defined as behavior that benefits others at a cost to oneself. The concept of altruistic punishment is widely used in a range of fields and considered to be a form of altruistic behavior in the sense that it provides a benefit to others at a cost to the punisher. Altruistic punishment has not been observed in animals other than humans and is thus considered to be a peculiar type of altruistic behavior that is unique to humans (Tei and Takahashi 2015; Jensen and Tomasello 2019).

The term "altruism", meanwhile, is, generally speaking, often thought of as a desire to help others, and includes the motivation to do so. This is the mental state of wanting to do something for the sake of someone else, even if doing so is costly or burdensome to some extent. In the words of social psychologist C. Daniel Batson, known for his work on altruism, "by altruism I mean a motivational state with the ultimate goal of increasing another's welfare" (Batson 2011, p. 20). In simple terms, we can think of altruism as the state of wanting to help someone else for the sake of that person's own benefit, regardless of whether there is any cost involved or whether the person actually benefits in the end.

A distinction between the former and latter is sometimes made, whereby altruism in the sense of conferring a benefit on others at a cost to oneself is known as **biological altruism**, while altruism when it refers to the actor's motivations, interests, and goals is known as **psychological altruism** (Morimoto and Tanaka 2016). For the remainder of this book, I will use the latter sense, which encapsulates the actor's feelings and motivations.

### Some usage notes on the term "cooperation"

The social behavior of animals including humans is, in general, classified into four types depending on whether the actor and others incur a cost or enjoy a benefit.

**Mutually beneficial behavior** is behavior that benefits both the individual doing the behavior as well as others. **Altruistic behavior** is that which benefits others at a cost to the individual. **Selfish behavior** is that which benefits the individual at a cost to others. **Spiteful behavior** is that which is costly to both the individual and others. In this book, I refer to the intersection of mutually beneficial behavior and altruistic behavior—behavior that benefits others regardless of whether or not it is costly to the individual—as **cooperative behavior**.

But because, as explained above, I am using "altruism" here in a sense that includes the purpose of and feelings behind a person's actions, in this book I will refer to altruistic behavior—that which benefits others at a cost to the individual—as **self-sacrificing cooperative behavior**.

### Empathy for pain

I discussed pain empathy in Sect. 4.3. The brain regions activated by your own physical pain include those involved in perceiving pain (such as the somatosensory cortex and the posterior insular cortex) and those implicated in making pain uncomfortable (such as the anterior insular cortex and anterior cingulate cortex). The regions of the brain activated by the physical pain of another person are the latter, those responsible for sensing the severity and discomfort of pain. These regions are implicated in how painful we perceive pain to be, which is to say the emotional or subjective sensation of pain.

The notion of the subjective sensation of pain might seem a little vague, so let's look at an example. People who have suffered damage to part of the cingulate cortex, one of the regions responsible for subjective pain, still feel pain just as they did before and are able to identify where the pain is and how much it hurts. But despite this, upon being physically injured, people with this sort of brain injury will report that the pain doesn't hurt or that it doesn't bother them very much.

On the other hand, people with damage to the somatosensory cortex, one of the regions responsible for pain perception, will display clear discomfort when receiving a painful stimulus, but not only are they are unable to describe the actual pain sensation—they cannot tell if the pain is coming from, say, heat or cold or a pinprick—they are also unable to identify the location of the pain. It's difficult to imagine what it must actually be like to have a condition like this, but cases such as these have helped us understand the role of these brain regions (Lieberman 2013b; Linden 2016).

### The sour aftertaste of punishment

In Sect. 4.3, I discussed how sweet it feels to punish wrongdoers. Subsequent research has revealed that the joy of punishment is a little more complicated than this simple observation might suggest.

Researchers studying the topic found that people derived greater satisfaction from punishing others when the punishment was not simply a response to a random bad event and when the offender understood that the punishment was a response to his bad behavior (Gollwitzer et al. 2011). Others have also reported that punishers want offenders to understand *why* (and sometimes *by whom*) they are being punished (Molnar et al. 2023). Another experimental study suggests that victims perceive justice as being served not through punishment itself but through changes in the offender (Hechler et al. 2023).

Other research, meanwhile, has indicated that even though punishers predicted beforehand that they would feel good about punishing a bad actor, they actually experience negative emotions after imposing the punishment (Carlsmith et al. 2008). While this may seem to contradict the research discussed so far, subsequent work has shown that people experience mixed feelings after administering punishment (Eadeh et al. 2017; Eder et al. 2021). It is possible that feelings of negative emotion after punishing someone have something to do with a deterrent mechanism that prevents punishment from becoming excessive, and a mechanism that, in turn, leads us to forgiveness.

Indeed, one study showed that punishment may elicit forgiveness by restoring a sense of justice (Strelan and van Prooijen 2013). We also know that a strategy of being tolerant and forgiving of cheating (or defection, as it may be called in different contexts in the literature) to an extent ultimately results in a higher overall benefit to the individual than a strategy of immediately punishing any cheater-like behavior, and that this can encourage cooperative behavior in society. These discussions surrounding the pleasure of punishment, subsequent complex emotions, and the potential for forgiveness seems to suggest that there may be a kind of rationality lurking beneath our emotions, which have often been held up as typical exemplars of irrationality (Kobayashi 2021).

**Field experiments on third-party punishment**

In Sect. 4.4, I indicated that costly third-party punishment occurs not only in the lab but also in field experiments. It must be noted, however, that the incidence of costly third-party punishment in these field experiments is only around 4–20%, which is significantly lower than in lab experiments.

In post-experiment surveys, the most commonly given reason for not exerting direct punishment (confronting a violator) was fear of retribution, with some 60–70% of respondents giving this answer (Balafoutas and Nikiforakis 2012, Balafoutas et al. 2014a). Work in this area has also shown that this fear of retribution may be stronger the greater the severity of the bad behavior being punished (Balafoutas et al. 2016). In a lab experiment, the incidence of costly third-party punishment dropped all the way from 50 to 10% when the threat of retaliation was present (Balafoutas et al. 2014a).

Extensive field research, however, indicates that punishment (or sanctioning) in human society is graduated (Ostrom 1990). Initial sanctions are surprisingly low, and could be described as being less like *wielding the whip* by actively reducing the offender's share of a resource and more like *withholding the carrot* by refusing to treat the offender kindly. And if that's the case, it does engender the possibility that punishment is not always being properly measured in experiments that look at whether people who have done wrong are being reprimanded or not.

**Issues of punishment behaviors**

The issue of punishment discussed in Chap. 4 leaves many questions unanswered. Even if punishment is deemed beneficial for maintaining societal order, there remain numerous aspects to consider around why people choose to engage in the act of punishing, which may not directly benefit them as individuals.

Insights from field research, for instance, have suggested that while punitive behavior is widespread, it is actually rare for people to directly administer punishment when they are the sole bearer of the cost (Baumard 2010; Guala 2012). People select the type of punishment to use depending on the situation and their relationship with the offender (Molho et al. 2020), and it is actually more common for people to employ punishments that do not impose a cost on the punisher, which can include punishments that obviate the need for face-to-face contact, such as rumors and gossip, as well as verbal teasing, admonition, or even social exclusion. Researchers have even reported that people prefer to impose punishment as a group (coordinated punishment) rather than as an individual (Molleman et al. 2019).

Moreover, evaluations of punishers are not always high (if the punisher enjoys a good reputation, then the mechanism of punishment could be explained by indirect reciprocity, which is described in the supplementary Section of Chap. 6).

According to research by Polly Wiessner, known for her detailed investigations, behaviors that involve frequent punishment are not highly favored, and instead, the ability to mediate is held in higher regard (Wiessner 2005). Indeed, the finding that mediation skills valued more is something that resonates with me when it comes to everyday life. We also have evidence that individuals of high prestige (those who command genuine respect) tend to be self-deprecating and humble and avoid showing aggression (Henrich 2017).

Punishers are not always held in high regard in experimental settings either. A punisher may potentially end up being trusted more (Jordan et al. 2016a), but is also liable to suffer negative consequences such as not being chosen as a future partner or being passed over for higher rewards (for a review, see Raihani & Bshary 2015a).

That said, the more problematic aspect of evaluating punishers lies in third-party punishment, since second-party punishment, being direct retaliation by the victim, could be explained by the benefit of protecting oneself in the future (Kobayashi 2021). For example, it has been reported that second-party punishers tend to receive a lower evaluation than third-party punishers (Mifune et al. 2020; Li and Mifune 2023).

However, while there was potential for third-party punishers to be rated highly, research indicated that people who helped the victims (third-party helpers) were allocated a greater reward (Raihani and Bshary 2015b) and were more highly valued (Dhaliwal et al. 2021) than third-party punishers. Even children seem to prefer people who help those who were treated unfairly over people who punish the wrongdoer (Lee & Warneken 2020). Field work has also shown that punishers are not particularly likely to receive help from others (Balafoutas et al. 2014b). There is still much that we do not know about the ultimate factors that drive the third-party punishment process.

And although it has for some time been thought that people who inflict costly punishment (altruistic punishment) are likely to behave fairly themselves, research has shown that this is not always the case (Yamagishi et al. 2012; Yamagishi et al. 2017; Hoeft and Mill 2017). For instance, researchers have reported that participants in the ultimatum game and other such games do not necessarily make fair offers when they switch into the proposer role. Punishers are a mixed bunch that includes not only those who seek to bring about fairness but also those who seek to harm an opponent who has regarded them as inferior (Yamagishi et al. 2012; Yamagishi et al. 2017).

In recent years, issues brought about by punishment have attracted increasing attention. For example, punishment could prompt retaliation (Herrmann et al. 2008; Li et al. 2018), it could trigger feuds (Nikiforakis et al. 2012), it could damage trust and hinder cooperation (Fehr & Rockenbach 2003), it could compromise morals (Gneezy and Rustichini 2000), and it could make people feel that others are only cooperating out of fear of punishment (Mulder et al. 2006). Other work shows that the threat of punishment itself may negatively affects people's ability to process information (Ballard et al. 2019).

Raihani, who has studied cooperation, states that "This inability to de-escalate conflicts makes punishment a dangerous, and potentially explosive, tool for trying to enforce cooperation" (Raihani 2021). People I have recently spoken with in various fields, such as the judiciary, healthcare, and government administration, all agree that punishment could exacerbate problems through the breakdown of trust and other factors. Indeed, there are reports suggesting that excessively harsh punishment causes inefficiencies (Kamei 2020), and that punishment may not necessarily enhance future gains for individuals or groups (Dreber et al. 2008, Rand et al. 2009).

The renowned anthropologist Dunbar says, "In my view, punishment is a poor way to create social cohesion; you cannot force people to be nice to each other." That's a view that I basically share. Is it not fair to say that punishment, at most, can only motivate ostensibly good behavior that falls just short of being punishable?

One of the key messages to come out of game theory, and of the theory of repeated games in particular, is that the most important thing in maintaining and building cooperative relationships is how well the punishments are designed. Yet when I ponder the findings discussed here, I start to wonder whether game theory isn't seriously wrong on this point. Naturally, I am not saying that punishment is unnecessary, but I also do not think it is punishment that motivates cooperation in the true sense—cooperation that goes beyond shallow, only superficially cooperative behavior.

A proper understanding of the essential nature of punishment will be crucial to understanding the mechanisms that produce cooperation. Research into the issue of punishment is currently thriving.

# Chapter 5
# Ingrained in the Brain

## We Detect, Remember, and Expose Cheaters

## 5.1 Powers of Cheater Detection

### Evolved abilities

Time for another quick recap. We have learned that Player 1, the proposer, is concerned with watching eyes and reputational effects, and is very sensitive to the idea of looking good (being seen as fair and just). We also know that Player 2, the responder, feels intense disdain and anger when intentionally treated unfairly and shows a strong desire to punish such treatment. Now that we have that groundwork established, it's time to ask just why it is that Player 1 is so concerned with looking good. That's the question we'll explore more deeply in this chapter.

Leda Cosmides and John Tooby, two pioneers of evolutionary psychology, argue that evolution has equipped our brains with some intriguing adaptations that are of relevance here. Our brains, it seems, are very adept at detecting cheaters (Cosmides 1989; Cosmides and Tooby 1989; Cosmides and Tooby 2005). To explore this claim, we'll first consider the following problem.

### The Wason selection task

The Wason selection task (also known as the four-card problem) is a well-known logic puzzle devised by cognitive psychologist Peter Cathcart Wason. Participants are given the task of determining whether a conditional rule of the form "If P, then Q" is violated.

Here's an example. Imagine you are presented with four cards, each with a letter on the front and a number on the back, arranged as shown in Fig. 5.1.



K. Kobayashi, *The Mystery of the Ultimatum Game*,
https://doi.org/10.1007/978-981-97-8979-5_5

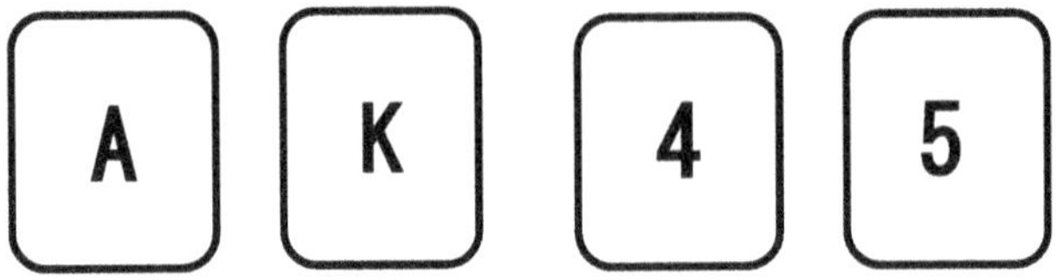


Rule: If a card has a vowel on one side, then it has an even number on the other side

**Fig. 5.1** The Wason selection task (abstract version)

Now, the rule is this: *If a card has a vowel on one side, then it has an even number on the other side.* Which card or cards do you need to turn over to determine whether the rule is true or false? This is the Wason selection task. Naturally, the puzzle requires that you avoid turning over any card unnecessarily. That is, you should only turn over the card or cards that you definitely need to turn over to arrive at a logically sound answer.

I hope that's all clear. While the problem itself is not terribly complicated, the implications of the rule can potentially take a little bit of thinking to sort out in your head. Figuring out the right answer is another matter entirely, however. Personally, I get a little flustered. I end up wanting to blurt out, "There's only four cards, right? Just turn them all over!" Typically, fewer than 10% of participants give the logically correct answer, and indeed, the success rates in past experiments range from 5% to at most 30% (Cosmides et al. 2010).

Okay, let's work through the answer. First, consider the K. This is a consonant, but a consonant can have either an even or an odd number on the other side and still be in conformance with the rule. That is, it doesn't matter what's on the other side of this card, and so there is no need to turn it over. Now consider the 4. This is an even number, so if there were a vowel on the back of the card, then the 4 on the front of the card would be in conformance with the rule. But then again, if there were a consonant on the back, then either an even or an odd number on the front would suffice. So either a vowel or a consonant on the back of this card would be fine, so there is no need to turn over the 4.

Now we move onto the A. As this is a vowel, anything other than an even number on the other side of the card would be in violation of the rule, so we need to check it. With the 5, the reasoning is similar. A vowel on the other side of the card would break the rule, so again, we have to check. And so the logically correct answer is "A and 5".

**Wish you were beer**

Look, I know at least some of you will still find this all a bit muddled after that explanation. But fret not. You're not alone. Truth be told, it does my head in too. So let's make it easier!

Take a look at Fig. 5.2. Say that on the front of each card appears someone's age and on the back of the card, that person's beverage. Here is the rule: *If a person is drinking beer, then the person must be over 19 years of age.* Which card or cards should you turn over to see if the rule is being followed?

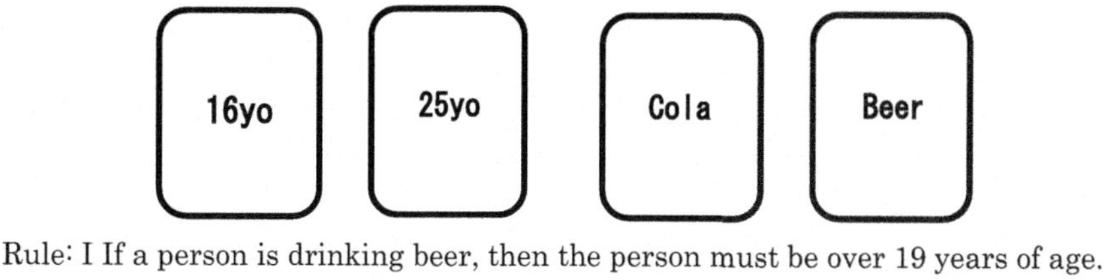


**Fig. 5.2** Wason selection task (alcohol version)

You need to figure out who's on the sauce despite being under 18 and thus too young to drink. Not only is the question easy to understand this time, I'd wager you've already figured it out! You will turn over "16yo" and "Beer". Let's check our answer, though.

The 25-year-old may drink whatever he or she wants, so there's no need to check! And there's no need to check who's drinking cola! You do need to check what the 16-year-old is drinking, and you do need to check how old the beer-drinker is. And so there it is. Turn over "16yo" and "Beer" and you're done and dusted! (Fig. 5.3).

**Fig. 5.3** Wason selection task (alcohol version)

**To catch a cheater**

Isn't that weird? The logical structure of the two questions we answered was identical. Yet with the alcohol version we just looked at, some 65–80% of experimental subjects are able to answer correctly, a ridiculously high success rate in light of the paltry figure for the abstract version we first looked at Cosmides et al. (2010). Follow-up work in Japan has produced similar findings too (Hasegawa and Hiraishi 2000). Why the disconnect?

Look at the alcohol version (*If a person is drinking beer, then he must be over 20 years old*). The problem here can be framed as the task of catching the little scallywag who, despite being under 18, is drinking and thus breaking the rules. That is, you need to figure out who's violating the rules of our society just to please themselves. The abstract version is full of numbers and letters and has no such emotive content, no mention of cheaters or miscreants or anything that might pique your social senses.

Now, when researchers tested a whole range of Wason selection task patterns, they found that the correct answer rate jumped when the question could be framed as a "look for cheaters" affair. In other words, we humans are good at catching miscreants, traitors, truants, and cheaters when we are told explicitly that that is our task. This ability is called **cheater detection**.

**Relatability and real-world grounding**

Hold up, though. Can we really be so easily convinced? The answer to the great beer and cola conundrum certainly was simple in the end. But is the explanation for this really something so elaborate as an innate human capacity for cheater detection? Does the ease with which the puzzle was solved in this instance not reflect the simple turn of phrase and the concrete, real-world nature of the rule?

Let's turn to another problem here (Cosmides 1989). Take a look at Fig. 5.4. On the front of each card is a destination, and on the back is the means of transport. Here's the rule: *If a person goes to Boston, then he takes the subway*. This experiment was conducted twice and involved 24 Harvard University undergraduates.

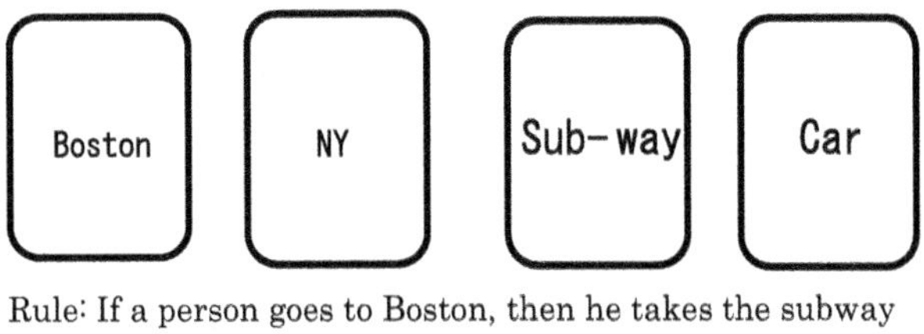


**Fig. 5.4** Wason selection task (a concrete and familiar scenario)

The task here is to figure out which card or cards you should turn over to determine whether the rule (*If a person goes to Boston, then he takes the subway*) is violated. There is no "look for cheaters" aspect to this scenario, but the situation and the rule are familiar and concrete (grounded in the real world).

Even so, you can be forgiven for concluding that the answer is not quite as easy to figure out as it might first seem. The correct answer is "Boston" and "Car". When the experiment was performed twice, only 46% and 38% of subjects, respectively, answered correctly. These figures are certainly not quite as high as those observed with the alcohol version. The correct answer rate was still higher than for the abstract version (letters and numbers), but the improvement was not as large as you might expect. All of this is not to say that making the rules concrete and the situation more relatable doesn't help, but that, apparently, is not enough on its own.

**Cheater detection? Concreteness?**

Let's move on to another problem.

The narrative this time is that the cassava root can only be harvested in certain areas, and people who live in areas where the cassava root grows customarily get tattoos on their faces. Because there is no means for distributing the product to other areas, only people in areas where the cassava root grows can eat it, so people in areas where the cassava root is eaten have tattoos on their faces. This yields the following rule: *If a man eats cassava root, then he must have a tattoo on his face.*[1]

Take a look at Fig. 5.5. Here, the front of each indicates whether a person eats cassava root, and the back of the card indicates whether the person has a face tattoo. Which card or cards should you turn over to figure out if any cards violate the rule?

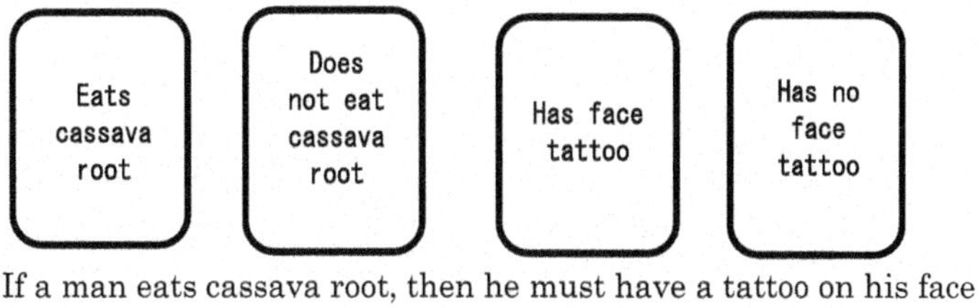


**Fig. 5.5** Wason selection task (a concrete but unfamiliar scenario)

Many people encountering this scenario probably have little, if any, idea about what the cassava root actually is, and the custom of tattooing one's face is most probably unfamiliar to most. So while this scenario presents a concrete rule, the situation is (quite likely) unfamiliar.

While this scenario may not be at all unintelligible, you may still feel a little muddled by all the details, such that the answer is not immediately obvious. The answer is "Eats cassava root" and "Has no face tattoo". Only around 20% of subjects answered this correctly, however, not a very high success rate (Cosmides and Tooby 1989).

### Cheater detection

Now then, let's frame the puzzle a little differently. Say you are told that cassava root is delectably delicious and highly nutritious. Everyone wants to eat it. But this wonder food is, unfortunately, quite rare. This makes it highly prized, so much so that only people with facial tattoos proving their bona fide tribe membership may eat it.

The cards have not changed, nor has the rule, and (were it so to begin with) the cassava root's status as a food and the custom of tattooing one's face as a sign of tribe membership remain quite unfamiliar concepts. That is, the situation and the rule remain unfamiliar in nature, just as they were before.

And yet those slight changes to the narrative were all it took to boost the correct answer rate to over 70%. Indeed, you also may have found the answer to be instantly obvious this time around. All you have to do is look for the cheaters, those miscreants who, with no right to do so, dare to eat the delicious, highly prized cassava root that

[1] Cassava is a woody shrub that grows in tropical and subtropical regions and is known for its starchy root tuber. Bubble tea fans may be interested to know that tapioca is produced from the cassava root.

everyone covets. The answer is, naturally, also unchanged: "Eats cassava root" and "Has no face tattoo".

A cheater is someone who enjoys a benefit without paying the proper price for it. In this scenario, the benefit of eating the delicious, highly prized cassava root must be paid for by paying the price (cost) of having your face tattooed, and so a cheater is someone who eats the cassava root (enjoys the benefit) without having the proper facial tattoo (without paying the price).

The first version of this scenario simply said that people in a certain region eat a food that can only be grown in that region, so there is no obvious benefit to gain or cost to pay, and thus no suggestion that you might need to catch anyone out. Yet, with these same cards and the same rule, a simple reframing of the question to introduce a "look for cheaters" element more than tripled the correct answer rate.

We've now looked at a few variations of the Wason selection task. Isn't it surprising how the implications of an unfamiliar situation and rule can suddenly become so clear when the problem is framed in terms of catching out cheaters, how effortlessly we figure out how to spot rule-breakers?

**Inadvertent errors**

We've learned, then, that humans are quite adept at catching out cheaters. And it turns out that intentional deceptions are more readily detected than inadvertent ones (Fiddick 2004). The experiments we will now look at involved 89 anthropology undergraduates at Harvard University.

The scenario here is about enrolling in children's swimming classes. It goes like this. If you want to enroll your child in the advanced class, you must pay an additional $50 because that class lasts longer. So the rule here is: *If a child is to be assigned to the advanced class, then that child must have paid an additional $50.* Take a look at Fig. 5.6. The front of each card tells you which class a child was enrolled in, and the back tells you whether an additional fee was paid or not.

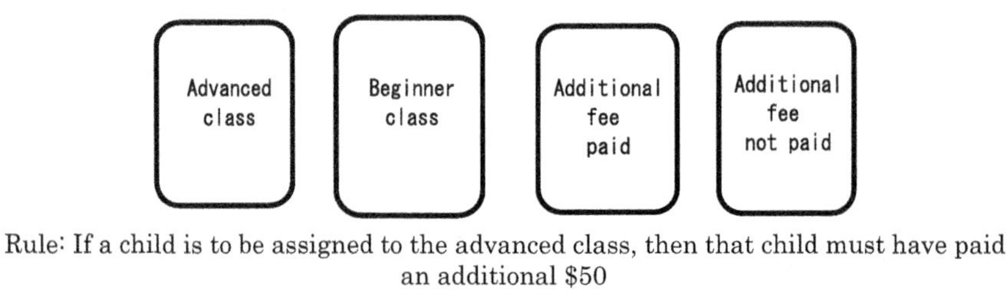


**Fig. 5.6** Wason selection task (swimming lessons and the additional fee)

Now, imagine you're told that an elderly lady will be completing the enrollment. But the elderly lady has been diagnosed with Alzheimer's disease, and while she has no intention of deceiving anyone about paying the correct fee, she sometimes ends up accidentally putting the child in the wrong class without paying.

Did you get the answer? This situation will possibly seem a little vague to some. And indeed, only around 45% of experimental subjects answered correctly in this case, not a terribly high success rate. The correct answer is "Advanced class" and "Additional fee not paid".

In this case, the elderly lady's illness can cause her to mistakenly enroll her child in the wrong class without paying the proper fee. In other words, she sometimes makes an inadvertent and entirely unintentional error. As we'll see next, however, when this narrative is changed to reframe the situation slightly, the results change dramatically, even though the underlying logic of the problem remains the same.

**Intentional cheating. . . unacceptable!**

In the new narrative, the child's father will be handling the enrollment. Said father is an incorrigible skinflint and known to bend the rules to save a few cents at any opportunity. He most certainly would love to avoid paying the additional $50 fee, and he will deliberately cheat by putting his child into the advanced class without paying that fee if he can get away with it. In other words, he is happy to intentionally break social rules for the prospect of an undeserved benefit.

Now then, do you not find yourself suitably enraged and ready to weed out this horrible human being? How dare he cheat! Isn't it surprisingly easy to figure out the answer in this case? The answer, of course, is again "Advanced class" and "Additional fee not paid". When the scenario was framed like this, the correct answer rate actually jumped to over 80%.

It is immediately obvious here that both the elderly lady and the father, having not paid the proper price (additional class fee) for the benefit gained (child enrolled in advanced class), are cheaters.

But the elderly lady made an inadvertent error because of her illness, while the father cheated intentionally. The upshot of all this is that with the same cards and the same rule, simply changing the narrative such that participants have an incentive to catch a cheater nearly doubles the correct answer rate. How did you find it? Wasn't the situation a lot clearer and the answer a lot more obvious when you knew the father was trying to cheat? We humans are extremely sensitive to intentional cheating, it seems (see Fig. 5.7).

**Fig. 5.7** Wason selection task (swimming classes)

**Hazard management scenarios**

Okay, we now know that we are extremely sensitive to intentional cheating. And in Sect. 4.1, we discussed evidence suggesting that humans are more sensitive to intentional unfairness than to unfairness per se. You might therefore be tempted to conclude that humans are sensitive to intentional deception. And yet, interestingly, this is not necessarily what's going on here.

There is another narrative pattern that is known to increase the correct answer rate—the **hazard management task**. With this pattern, the rule is worded in a way that requires people to take safety precautions when engaging in dangerous activities. Examples include: *If you are working at height, then you must wear a hard hat*; and: *If you work with poison, then you must wear rubber gloves*. And with hazard management scenarios like this, people were able to detect both intentional and inadvertent deceptions at pretty much equally high rates.

Let's look at a concrete example by revisiting the children's swimming class enrollment problem (Fiddick 2004). Take a look at Fig. 5.8. The narrative here is that only experienced swimmers may join the advanced children's class because the advanced class is too dangerous for beginners who are not yet adept at swimming as they get fatigued too easily, which presents a real risk of drowning. So the rule is: *If a child is to be assigned to the advanced class, then that child must be an experienced swimmer*.

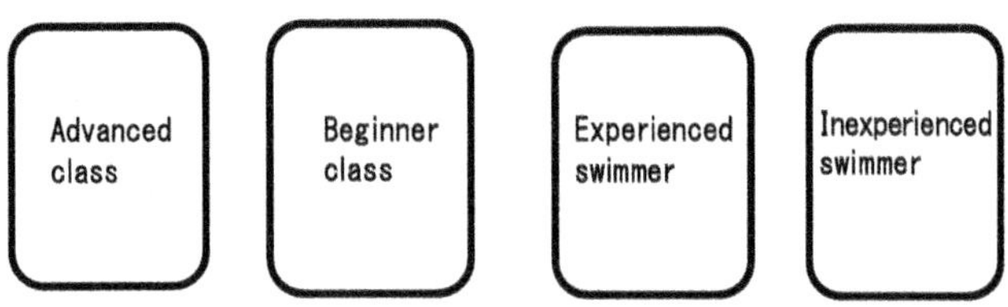


Rule: If a child is to be assigned to the advanced class, then that child must be an experienced swimmer

**Fig. 5.8** Wason selection task (swimming lessons and the advanced class)

This rule says that if the child is going to enroll in the advanced class and thus undertake a potentially dangerous activity, then to mitigate the risk posed by that danger, the child must be a proficient swimmer. So you can see that this is most certainly a hazard management task.

**Inadvertent errors and deliberate cheating**

Let's consider this hazard management situation both in the presence and absence of intent. Suppose you are told that the elderly lady who suffers from Alzheimer's disease could potentially enroll the child in the wrong class by mistake. This is the inadvertent error scenario. On the other hand, suppose you are told that the father yearns to see his son excel and is a little ambitious about his child's prospects, and therefore decides to flout the rule and enroll the boy in the advanced class.

Both of these narratives adopt a very similar form and rule to the previously discussed versions in which you had to catch the cheater out. But surprisingly, the results were completely different in this setup.

In these hazard management scenarios, over 80% of participants answered correctly when thinking about the intentionally cheating father, a high figure that, perhaps unsurprisingly, mirrors the outcome under the previous version of the father scenario. In the case of the inadvertently erroneous elderly lady, however, a tad over 79% of participants got the answer right—quite a change. That is, the correct answer rate was high at around 80% for both scenarios. And no doubt this makes sense to you, because both of these scenarios are a little worrying. In both cases you're probably wondering if the improperly enrolled child will be okay. I'm sure the answer is obvious by this point: "Advanced class" and "Inexperienced swimmer".

With the earlier cheater detection scenarios, participants more readily detected cases of intentional cheating than cases of inadvertent errors. But with the hazard management scenarios, participants were able to figure out whether the child was in danger at equally high rates under both the intentional cheating and inadvertent error narratives. What's more, when presented with hazard management scenarios like this, participants seem to be more successful at detecting when their own child is in danger than at detecting when someone else's child is in danger (Thompson et al. 2015).

**The discerning brain**

We've labeled people who break the rules for their own benefit free riders. And if you really think about it, inadvertent (and thus unintentional) errors do not a free rider make. Plus, by constantly pulling someone up on their inadvertent errors, you run the risk of fermenting discord. Perhaps it's best, then, not to be so persnickety. Perhaps it's best to simply not notice every little inadvertent error someone makes.

Deliberate cheaters, on the other hand, are the very model of a free rider. And once they have a taste for some nefarious behavior, there's no guarantee they won't do it again, over and over. This is the sort of cheating that we most certainly cannot let slip. All that aside, a child in danger is a child in danger—if the child is a beginner in the advanced class, say, the question of whether the child has been placed there inadvertently or deliberately does not matter to us. We have the ability to unconsciously distinguish between logically identical scenarios like this, a finding that makes one marvel at the mysterious power of the brain.[2]

When we conducted a large-scale experiment involving thousands of participants, replacing the Wason selection task with a financial context, the results discussed here

[2] Strictly speaking, though, rather than making a distinction, the brain seems to treat them as completely different things. Work has shown, for example, that different regions of the brain are activated when solving cheater detection and hazard management problems (Fiddick et al. 2005). In fact, when people with brain damage were asked to solve these sorts of problems, they were able to do one but not the other (Stone et al. 2002). In terms of emotion, research has also been shown that cheater detection situations are associated with anger, whereas hazard management situations are associated with fear (Fiddick 2004). Given these findings, it is thought that the brain treats the two types of situations as completely different things.

were clearly replicated: only intentional cheating led to an increase in accuracy rates when compared with cases of accidental mistakes.

Then, using the finding of Cosmides et al. (2010) (discussed later) that the correct answer rate increases when there is a benefit to the cheater, we show that people are able to make more prudent decisions about risk when prompted to be aware of the obvious fact that financial institutions are businesses and thus pursue financial gain (Kobayashi and Motonishi, mimeo).

It has also been shown that women are somewhat more sensitive to intentional cheating than men, and higher cognitive ability is associated with higher rates of response accuracy. An interesting finding here is that sensitivity to intentional cheating is not dependent on degree of cognitive ability. That is, by making people aware of the benefits that others stand to gain, participants with comparatively lower cognitive ability were able to reach a similar level of performance to those with high cognitive ability. Sensitivity to cheating seems to have little to do with so-called cognitive ability.

Intriguingly, when we looked at the data on the number of cards participants turned over, we found that in both the case of inadvertent errors and of intentional cheating, the number of cards turned over increased by the about same amount compared with the neutral case. Given that the correct answer rate for inadvertent errors was pretty much the same as for the neutral case, this means that participants were unable to arrive at the correct answer even after turning over more cards. Perhaps it is that in the case of advertent errors, while people do feel some sort of vague concern, this does not readily lead them to the right answer.

**I know when I'm being cheated**

Let's look at one more experiment that will take us a little deeper into what our cheater detection capabilities mean (Gigerenzer and Hug 1992). The experiment involved 93 students aged 19 to 35 at the University of Konstanz in Germany.

Say you are the owner of a small business and have four employees. You have a rule saying that to be eligible to receive a pension from the business, an employee must have worked there for over 10 years. So the rule is: *If a previous employee gets a pension from a firm, then that person must have worked for the firm for at least ten years.*

Take a look at Fig. 5.9. The cards tell you how long each employee has worked and whether or not the employee is receiving a pension. The cost of a pension must of course be borne by the business. Now, which card or cards should you turn over to find out if someone is breaking the rule?

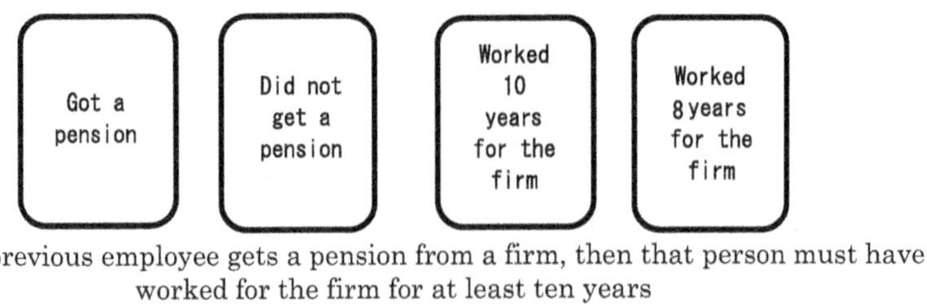


**Fig. 5.9** Wason selection task (pension eligibility)

So, which cards did you choose? To be eligible to receive a pension from the business, you must have worked there for at least 10 years. No doubt many readers will have answered that you should turn over "Worked 8 years for the firm" and "Got a pension" in order to figure out if anyone is cheating by receiving a pension without having met this criterion.

Now then, suppose you are asked to think about this same problem from the perspective of a labor union representative. That is, you are tasked with the important job of protecting the rights of workers at this business. The cards presented to you tell you the length of service and pension status of your former colleagues, whose rights you are sworn to protect. Now which cards do you want to turn over?

Surprisingly, when presented with this framing, more people chose "Worked 10 years for the firm" and "Did not get a pension". But as you no doubt immediately realize, this is the exact inversion of the cards chosen under the previous version. The cards are the same and the rule has not changed, and yet a simple shift in perspective is enough to elicit the exact opposite answer. What is going on here?

From the perspective of the business owner, as in the initial framing of this problem, a cheater is someone who is receiving a pension despite not having worked for the firm for over 10 years. You need to find out if anyone is sly enough to take a pension without being eligible for it, so many people will have accordingly zeroed in on the "Worked 8 years for the firm" and "Got a pension" cards.

From the perspective of a worker, however, being cheated would mean not receiving a pension despite having worked for the firm for more than 10 years. To figure out if any workers are being shortchanged and not receiving a pension despite having worked the requisite number of years, it's perhaps natural to want to choose "Worked 10 years for the firm" and "Did not get a pension".

So the definition of cheating in this scenario depends on your perspective. Because participants were trying to detect cheaters from their own perspective, or in other words, trying to figure out if they were being treated unfairly, participants given opposing perspectives chose the opposite cards.

To flip this finding around, we can say that people are somewhat oblivions to their own improprieties. We are quick to spot cheating by others, but we fail to notice when we ourselves cheat. We call ourselves *Homo sapiens*—wise humans. And so to think that the capacity for this sort of dualistic behavior is innate within us is, I think, a bit confronting.

**Social rank and cheating**

Research has also shown that we are better at detecting cheating by individuals with a lower social rank than ourselves than we are at detecting cheating by higher-ranked individuals (Cummins 1999). The experiment here involved 160 students from California State University-Sacramento and the University of California-Davis.

In this scenario, participants had to determine whether someone had violated the rules of the college dormitory. The experiment looked at what happened when participants were assigned different social ranks. In one setup, they were asked to take the perspective of the dorm's student leader, a high-ranking position, and determine whether a dorm resident, a low-ranking position, had followed the rules or not. In the other, they were asked to take the perspective of a (low-ranking) resident and figure out if the (high-ranking) student leader had upheld the rules or not.

Take a look at Fig. 5.10. When participants were asked to detect cheating in equal-ranked individuals (equally high-ranking or equally low-ranking individuals) or in higher-ranking individuals than themselves, the correct answer rate was only around 15–20%, not all that high. But when participants took a high-ranking perspective and looked for cheating in lower-ranking individuals, the success rate jumped to 65%. As the graph makes plain, this is more than three times the figure in the first case.

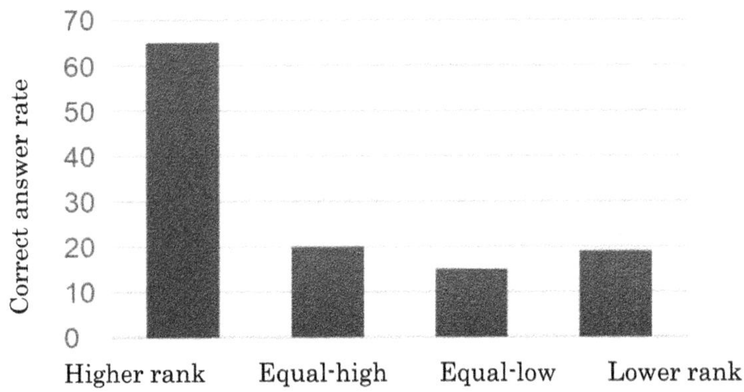


**Fig. 5.10** Social rank and correct answer rates. *Source* Based on Fig. 5.1 in Cummins 1999

Even if you discover cheating by a high-ranking individual, calling it out may not be so easy and, in fact, may simply draw more attention to you. But perhaps it is that overlooking or failing to detect cheating by lower-ranking individuals would be a serious blemish on your record that could jeopardize your own rank.

This finding that differences in relative social status have a bearing on how easy it is to detect cheaters is intriguing. Recall that in hazard management situations, too, researchers have reported that success rates are higher when people think of it being their own child who is in danger (Thompson et al. 2015).

One of the most well-known cognitive biases in behavioral economics is the anomaly of loss aversion: humans are more sensitive to losing something than to gaining something of equivalent value. I now think that the findings discussed here, including the pension problem and the social rank problem, together with this propensity for loss aversion may indicate that we humans are sensitive to cheating when that cheating is detrimental to ourselves.

### Catch the cheater!—an evolved ability

We've now covered many variations of the Wason selection task. I wonder if you're as utterly amazed as I am at how changing the situation ever so slightly can so dramatically affect how easy it is to see what's going on and what the correct course of action is.

We also know that this tendency to detect cheating is present in many cultures. Take, for example, the Shiwiar hunter–horticulturalists, who live deep in the Amazon. They receive no Western-style schooling, and they do not read or write, but they achieved success rates over 80% when presented with the Wason selection task. This was in fact about a percentage point higher than that of Harvard University students when presented with the same type of problem (Sugiyama et al. 2002). Cheater detection is also observed in children around the ages of 3–4, and it does not seem to have much to do with experience or familiarity (Cummins 1996; Harris et al. 2001). In light of these findings, Cosmides and colleagues have argued that the ability to detect cheating and see through deceit is something that evolution has bestowed upon our brains so as to protect us from cheaters (Cosmides and Tooby 2005).

Cosmides and colleagues later showed that we are not all that good at detecting cheaters when the cheating is accidental, when it does not benefit the cheater, or when the situation would make cheating difficult (Cosmides et al. 2010). In other words, it is easier for us to spot cheaters when there is more likely to be a possibility of cheating, when the cheating is more likely to be intentional, and when the cheater is likely to benefit from cheating.[3] While our radar is sensitive to intentional cheating for the purpose of gain, that sensitivity seems to fade when we feel that cheating is absolutely impossible (The end-of-chapter supplement has more information about the Wason selection task.).

## 5.2 Remembering and Exposing Cheaters

### Once a cheater, always a cheater

So we know that people have an innate ability to spot cheaters. We have another ability along these lines too: we don't forget a cheater.

The first experimental work on how we remember cheaters was done by Linda Mealey and colleagues (Mealey et al. 1996). The subjects were shown photos of people along with a fictional description that included the person's history of cheating. Subjects readily remembered the faces of people labeled as cheaters even after some time had elapsed. Similar results were obtained with Japanese subjects as well (Oda 1997). Mealey's work has been criticized, however, with other researchers falling to reproduce the findings, and pointing out that simply remembering whether you have

---

[3] Researchers have studied what happens in the dictator game when the ability to request a payoff is introduced (Yamamori et al. 2008, 2010). That is, Player 1 is asked to distribute a particular amount to Player 2.

This work found that Player 1 reacts differently when Player 2 makes the request ("I would like this much") as opposed to the request coming from a third-party (Player 3) with no direct interest in the interaction ("Please give this much to Player 2"). Even when the amount requested was the same, Player 1 was less accommodative of requests coming from Player 2 than of those coming from Player 3. While it may be too much of a stretch to say that making a request is cheating, this may have some association with the finding that people are more sensitive to cheating that benefits the cheater.

seen someone's face before does not mean you know or remember whether they are a cheater or trustworthy (Barclay and Lalumière 2006; Mehl and Buchner 2008).

Meanwhile, other work has shown that the faces of cheaters are rated more important to remember than the faces of cooperators, that people look longer at the faces of cheaters than at the faces of cooperators, and that people are more likely to remember cheaters and information about their transgression (Chiappe et al. 2004).

Subsequent work also found that source memory for cheaters (i.e., memory for the context in which a person was encountered) was better than source memory for other types of people (Buchner et al. 2009; Bell and Buchner 2009). We humans also seem to remember gossip information about a person better than we remember neutral information about the person or trivia-like information (Alicart et al. 2020), and we remember people who have acted solely in their own self-interest (failed to share) and rescind trust in such people (Oda and Nakajima 2020). Researchers have also found that we pay more attention to the faces of people who fail to cooperate than to the faces of those who do, which is to say that we have an **attention bias** toward uncooperative people (Vanneste et al. 2007), and that when two different photos are presented to each eye (binocular rivalry), we look longer at faces associated with negative gossip (Anderson et al. 2011).

We also better remember the faces of cheaters when the cheating is more damaging to us (Bell et al. 2014), and children more readily remember the faces of mean children than those of nice children (Kinzler and Shutts 2008). Think about it. When we're watching a movie or TV series, the villain's presence sometimes overshadows that of the hero in our minds, and this may have something to do with the tendencies we're discussing here.

This ability to better remember who has cheated and how is also thought to be an innate feature our brains have evolved to protect us from cheaters.

Incidentally, research also indicates that we readily remember not only cheaters but also people with high social status (Ratcliff et al. 2011). We generally want little to do with cheaters, and we often want to stay under the radar of high-status individuals, and so we remember who these people are. We unconsciously and quite readily remember people around whom we must tread with caution. Findings like these may make you think that just living your life is a bit daunting at times (Fig. 5.11).

**Fig. 5.11** We don't forget a cheater

### Gossip—sordid details following

Not only are cheaters quickly discovered, they are also unlikely to be forgotten. And cheating behavior quickly sows the seeds of gossip.[4]

People in general, it seems, view gossip and gossipers negatively (Haidt 2006), but research by the prominent anthropologist R. I. M. Dunbar indicates that around two-thirds of human conversation time is devoted to social topics, most of which can generally be labeled gossip (Dunbar 2004).[5] The psychologist Nicholas Emler has reported similar findings based on his studies of gossip (Emler 1994).[6]

Gossip is known to be a uniquely human phenomenon, and it is so prevalent across human societies that it also appears among the human universals (Brown 1991). Research shows that social information about person-to-person relationships tends to be remembered more accurately and in greater volumes than non-social information (Mesoudi et al. 2006; Redhead and Dunbar 2013). And humans prioritize social information over non-social information (Dunbar 2014).

Work by Polly Wiessner, who has studied the Ju/'hoansi people of northwestern Botswana for over 30 years, indicates that nearly 60% of conversations among the Ju/'hoansi are negative and involve criticism of a peer. Only 7% of conversations involved praise (Wiessner 2005). The famous anthropologist Christopher Boehm, who we also encountered earlier, had this to say about one society he visited during his fieldwork on the world's hunter-gatherer societies: "In the tightly networked settlement where we were living, it was all too apparent that the preponderance of this 'talking' was focused negatively on individuals suspected of deviant behavior" (Boehm, p. 243).[7]

---

[4] In the literature, "gossip" most commonly refers to any sort of talk about a third-party, whether the content is good, bad, or neutral (Foster 2004). In other words, it's anything said about a person behind their back and is not limited to negative rumors.

[5] It has been reported that people who engage in gossip are perceived as less moral and less competent, but more sociable compared to those who do not (Zhang et al. 2024). Furthermore, gossiping—whether it involves general, neutral, and sometimes even positive contents—could harm one's professional reputation (Gordon & Schweitzer 2024). However, some studies suggest that the impact of gossip on the gossiper's reputation may vary depending on whether the gossip is "positive" or "negative" (Lian et al. 2022). Moreover, research has also found that when women share negative gossip out of "concern" for others, it does not necessarily tarnish their reputation (Reynolds et al. 2025).

[6] Research suggests that even children view peddlers of negative gossip in a poor light (Shinohara et al. 2020).

[7] Research has also shown that the more disgusting a story, the more people want to pass it along to others (Heath et al. 2001); that threat-related information (rumors that convey information about potential danger) spreads faster than information that is negative and information that suggests potential benefits, and that people want to know threat-related information (Blaine and Boyer 2018); and that task-irrelevant threat-related information is hard to ignore (Zsidó et al. 2022).

Work has also shown that the negativity of a tweet (back when it was called Twitter) increases the probability of retweeting (Jiménez-Zafra et al. 2021), that when people are scrolling through online feeds, they pay more attention (according to dwell times measured via eye-tracking) to information that has a negative valence (Kohout et al. 2023), that the more negative words there are in a news headline, the greater the click-through rate (Robertson et al. 2023), false news spreads faster, deeper, and broader than the truth (Vosoughi et al. 2018). Moreover, it seems that it is humans, rather than

The famous social psychologist Jonathan Haidt notes the following when discussing a questionnaire survey of college students: "Among our main findings: Gossip is overwhelmingly critical, and it is primarily about the moral and social violations of others" (Haidt 2006, p. 54). Apparently, stories about others' good deeds were only about a tenth as common as stories about others' social and moral transgressions. In a survey covering a wider range of ages (18–68 years), participants reported learning about immoral acts more than twice as frequently as moral acts (Hofmann et al. 2014).[8] Another study found that when children report the behavior of classmates to a third party, tattling (reports of norm violations) represents the majority of those reports (Ingram and Bering 2010). Note that this tattling was usually truthful.[9] Should these findings indicate to us humans that we are spending prodigious amounts of time preoccupied, and perhaps even amusing ourselves, with negative gossip about the foibles and transgressions of others[10]?

In one study, researchers even found that our brain's reward system (striatum) is activated when we hear negative gossip about celebrities (Peng et al. 2015), and more recent work also produced similar findings (Alicart et al. 2020). Recall that the reward system is the brain region that lights up when we eat something delicious, and perhaps this has something to do with why gossip is often described as juicy.

---

automated spreading bots, who are spreading such fake news. This is because, while bots spread fake and non-fake news equally, humans tend to favor the spread of fake news more.

Fake news has become a major social issue, as you know, and while typical countermeasures like fact-checking and media reporting on falsehoods do seem to have some impact, they also appear to foster skepticism toward 'true' news (Hoes et al. 2024). Humans are indeed complicated creatures.

[8] In this survey, participants were also asked to report acts that they had committed, were the target of, or witnessed directly.

Moral and immoral acts had fairly similar frequencies for acts that participants reported being the target of or witnessing, but when it came to their own actions, participants reported committing about twice as many moral acts as immoral acts. That is, they apparently felt that they themselves behaved much better than others.

[9] In another experiment, researchers found that although lies do not seem to account for the majority of gossip, lies are actually about twice as frequent when people are gossiping about competitive rivals (Peters and Fonseca 2020).

[10] Others report a much lower frequency of negative gossip, however (Dunbar 1996).

And there are also data indicating that the content of daytime and nighttime conversations is quite different (Wiessner 2014). Wiesner's research indicates that daytime conversation is more functional, while nighttime conversation is more storytelling and socially oriented.

Others have reported that gossip between females is more negative than gossip between females and males or gossip between males (Hess and Hagen 2006). As mentioned in a brief aside in Sect. 4.3, women tend to use indirect aggression when attacking others, so the sort of negative or critical gossip observed in this study is possibly a means of making personal attacks.

Also, as discussed in the main text, a study on the way gossip changes within groups showed that when a problem arose within a particular group, both gossip critical of the person seen as the cause of the problem and gossip casting others in a positive light was observed (Kniffin and Wilson 2005).

This research tells us that the dataset collected, the way it is interpreted, and so forth can change greatly depending on how the researchers define terms like "gossip" and "critical gossip", and by when, where, and with respect to whom the data are collected.

Indeed, the oldest known piece of gossip, as far as we know, concerns an influential Mesopotamian man having an affair with a married woman (Wilkes 2002). This story was found inscribed on a cuneiform tablet dated to around 1500 BCE. Perhaps human behavior has not changed all that much since ancient times.

**Gossip—a policeman and a teacher**

Boehm points out that "It's by adding up information [obtained through gossip] that social deviants are identified and people can unite to cope with them" (Boehm 2012, p. 240). Gossip, he says, is what makes it possible for everyone to share information on who the deviants are and, thereby, what makes it possible for us to deal with such people as a united group instead of merely as individuals.

Research by Wiessner, whom I mentioned earlier, also indicates that gossip is more pervasive when the group faces a problem that it needs to deal with collectively. Research looking at how the content of gossip changes within a group also indicates that when the group faces a problem, the volume of gossip increases sharply, especially critical gossip about the person seen as the cause of the problem (Kniffin and Wilson 2005). And when the problem is resolved, gossip suddenly drops off.[11]

Emler, whom I also mentioned earlier, notes that gossip is a reputation management tool that enables us to know about, for instance, our own reputation and that of others (Emler 1994). And indeed, Haidt observes that "Many species reciprocate, but only humans gossip, and much of what we gossip about is the value of other people as partners for reciprocal relationships" (Haidt 2006, p. 55). In other words, gossip about someone else's bad behavior helps us figure out who we can trust as partners. Haidt also notes that the reputations we form through gossip are a key social mechanism that discourages undesirable behavior that could elicit gossip (and compromise our reputation).

"Gossip is a policeman and a teacher," says Haidt (2006, p. 55). That is, gossip and reputation make sure that bad actors get dealt with, and they teach us what sort of behaviors are regarded as undesirable.

Indeed, reputation has been shown to encourage cooperative behavior (Milinski et al. 2002). As discussed in Sect. 3.2, we humans are extremely concerned about what others think of us, and when we receive praise, this activates our brain's reward system and we feel joy. In Sect. 3.3, we also learned that being ostracized from a group is painful. Research has shown that gossip influences our decisions about who to interact with, and that we avoid consorting with people whose character has been compromised by negative gossip (Feinberg et al. 2014). The potential for people to gossip about you makes Player 1 present fairer proposals to Player 2 in the ultimatum game (Samahita 2017) and encourages cooperative behavior (Beersma and Van Kleef 2011; Fehr and Sutter 2019). Research also indicates that people are willing to incur substantial costs to protect their reputations (Vonasch et al. 2018) (Fig. 5.12).

---

[11] While gossip critical of the person seen as the cause of the problem increased sharply, positive gossip praising others also increased at the same time, and once the problem was resolved, both types of gossip fell away sharply. The researchers argue that gossip is not only a means of attaching bad actors but also a means of encouraging good members of the group and of affirming that "we're alright".

**Fig. 5.12** The power of gossip

**The power of gossip**

Haidt (2006, p. 54) also notes that "When people pass along high-quality ('juicy') gossip, they feel more powerful, they have a better shared sense of what is right and what's wrong, and they feel more closely connected to their gossip partners."[12] Experiments have also shown that gossip need not necessarily be accurate (Fonseca and Peters 2018), that gossip has a strong influence on people's behavior even when they have access to the original (accurate) information on which the gossip is based, and that people are not really all that concerned with the credibility of the information (Sommerfeld et al. 2007). These findings may seem all too familiar when you think about life in the real world. The power of gossip is something to respect, and perhaps even something to be a little afraid of.

Put simply, information about people's social violations and transgressions is quickly transmitted throughout the group so that everyone knows (and talks) about it. That's the power of gossip.[13]

[12] Research has also shown that gossiping may increase the hormone **oxytocin** (Brondino et al. 2017). Oxytocin has been called the "trust" hormone and is associated with trust, love, and the feeling of connection (Zak 2012). In this research, oxytocin did not increase in the case of neutral conversations and emotional non-gossip conversations; it only increased in the case of gossip.

The hormone **cortisol**, meanwhile, which has been associated with stress, decreased by a similar amount across all three types of conversations. So the act of conversation itself seems to be useful in relieving stress, but gossip may be the only type of conversation (of the types studied) that evokes feelings of connection with others. Hence, one function of gossip may be to strengthen interpersonal connections.

[13] Renowned anthropologist Robin Dunbar argues that "language evolved to facilitate the bonding of social groups, and that it mainly achieves this aim by permitting the exchange of socially relevant information" (Dunbar 1996, p. 174). In other words, language evolved so that we could share information about who gets along with whom and who doesn't, who did that awful thing the other day, and so forth. In Dunbar's words, "language evolved to allow us to gossip" (Dunbar 1996, p. 113). "In short," he also says, "gossip is what makes human society as we know it possible" (Dunbar 2004). The interested reader is directed to Dunbar's excellent book *Grooming, Gossip, and the Evolution of Language* (Dunbar 2016).

## 5.3 Error Management Theory

### Human society—we detect, remember, and expose cheaters

The discussion so far in a nutshell is as follows.

We humans have long had basic empathy for each other, and we have lived in groups because this is necessary for our survival. Unmitigated empathy for the pain of others presented a potential problem, however. It would hinder our ability to punish any free riders in our midst, a situation that, if allowed to continue indefinitely, would result in everyone's demise.

So we developed a sensitivity to unfairness, and feelings of strong dislike and anger toward intentional unfairness in particular. Not only were we thus emancipated from feelings of empathy for selfish free riders, we also developed the ability to enjoy punishing cheaters, whose suffering evokes joy and pleasure within us. And this ability to take joy and pleasure in the suffering of free riders was extended to all members of the group, not just those with a direct vested interest in the punishment interaction.

Moreover, we humans are extremely sensitive to (we quickly detect) cheating, and intentional cheating is particularly easy for us to spot. We can also spread information about such behavior rapidly throughout our group in the form of gossip. And we have a long memory for the faces of cheaters and other relevant information about them. We are adept at detecting, remembering, and exposing cheaters.

Realizing just how risky it would be to behave in a way that might be perceived by others as cheating in a society like this requires very little foresight. With that in mind, it would seem that from a long-term perspective, it is in your interest to show some consideration for others, instead of acting selfishly in pursuit of immediate, short-term gains. The mechanisms that enable us to consider others' interests are something that evolved within us to ensure our survival.

### A watching-eyes redux—which errors are more serious?

Recall the discussion of watching-eyes in Sect. 3.2. We humans are naturally sensitive to the watching eyes of real people, and we are even affected by mere images of watching eyes or eyespots printed on a poster or banner or the like. There were no photos in the ancient landscape of our evolution. Our brains naturally came to react to the presence of eyes of any sort as if someone (or something) was watching.

This explanation of human sensitivity to watching eyes comes up a lot, but there's more to the story here. I think the question of what types of mistakes are more likely to be fatal is actually more important.

We can make two types of errors when determining whether someone is watching us. We might think no one is there despite a watcher being present (false negative), or we might think someone is there when the coast is actually clear (false positive).[14]

[14] Anyone with some sort of background in statistics may remember these as type I and type II errors. If statistics isn't your thing, these terms are not important here, so forget I mentioned it!

In simple terms, thinking that someone is watching when no one is really there could result in you missing out on a good opportunity to do as you please. This could, admittedly, be a little disappointing.

But the mistake of believing no one is there when someone is actually watching could be a fatal one. If you nonchalantly, perhaps even gleefully, do something selfish under the false belief that the coast is clear despite there being a witness to your actions, you may find yourself labeled a cheater in very short order. Thinking back on the discussion so far, I'm sure it's pretty obvious which type of error is more serious.

## Sensitivity to crucial errors

This idea that different kinds of errors are more serious, or more costly, comes from **error management theory** (Haselton and Buss 2000). For our purposes, however, it may be easier to think of this in terms of the distinctively named **smoke detector principle** (Nesse 2005).

Smoke detectors are supposed to warn us if a fire has started. An overly sensitive one will sound the alarm at the figurative drop of a hat, needlessly startling everyone in the vicinity. So the cost of errors produced by a highly sensitive smoke alarm is the annoyance of false alarms. The error cost of reducing the sensitivity, however, is the failure to detect actual fires. Which is worse, the annoyance of a false alarm or the danger of an undetected fire? False alarms can be a bit of a pain, but knowing that an undetected fire could have much greater repercussions, it is in our interest to give smoke detectors a decent level of sensitivity.

Turning up the sensitivity will of course result in more small errors in the form of false alarms. But this can help us avoid the potentially fatal error of missing a real fire. As Kenrick and Griskevicius put it, "The smoke detector principle leads people to make this small error to avoid the much costlier mistake of getting into an accident without wearing a seat belt" (Kenrick and Griskevicius 2013, p. 81).

Our extreme sensitivity to watching eyes has been explained as an artifact of the brain mistaking fake eyes, which were not present in our evolutionary environment, for real eyes. We can go a step further, though. I personally think that this extreme sensitivity to eyes is likely an adaptation we acquired through evolution because the error cost is low and, hence, the potential benefit is high.

If it helps you to avoid more serious errors, then making a whole bunch of trivial mistakes may actually be more beneficial in the long run. Our brains are not focused on acquiring correct information per se, but rather on being able to make the "right" decisions (Fig. 5.13).

**Fig. 5.13** The smoke detector principle

**Error management theory in the ultimatum game**

The error management theory approach can also be applied to the ultimatum game and the dictator game as a whole. Let's recap those games. You are Player 1. A researcher appears, gives you $10, and asks how much you will offer to Player 2, a complete stranger sitting in the next room. Will you offer up some of the cash, or will you refuse to part with any of it? You must decide whether you will act in a way likely to be perceived as kind to strangers or in a way likely to be perceived as entirely selfish.

The error cost of behaving in a manner likely to be perceived as kind to others is losing out on the money you could have kept for yourself. That is a bit of a loss, to be sure. But the error cost of behaving in a manner likely to be perceived as entirely selfish is the risk of others thinking something like "What a selfish pig!".

Which is worse? Given the discussion so far, I hope it's evident that the risk of being labeled a selfish actor by members of your group is far more serious. That being the case, reducing that risk by offering some money to the other player and thus choosing to behave in a way seen as kind to others is likely to be more beneficial to you in the long run. The ultimatum game as a whole has a giant smoke detector built in. To me, error management theory is one of the fundamental ideas that gives us an important perspective when thinking about the biases of human cognition.[15] This of course means, though, that actions are being chosen in an effort to maximize some variable.

**A brief aside: Error management theory application 1—agents and agency**

In the main text, we discussed the special sensitivity that we humans have to eyes and faces, but it is also argued that we are sensitive to the presence of other "agents" (any entity, such as an animal, that has agency).

When, say, a blade of grass moves or a nut falls from a tree, we must quickly determine whether the event is simply due to natural phenomena, such as wind, or the sign of a hidden predator. We can of course make two types of error here: detect an agent when none is present (false positive) or fail to detect an agent when one

[15] See Yamagishi et al. (2007) for an application of error management theory ideas to the prisoner's dilemma game.

really is lurking (false negative). In light of the danger an undetected predator would pose, it's obvious which of these carries the greatest error cost—the false negative. If that's the case, the brain should increase the sensitivity of its agent detector module so that we can avoid the false negative error (failing to detect an agent when one is present). This would of course make us more likely to detect an agent when none is there, such that we often see ghosts in the shadows for instance. This cognitive tendency of our brain to attribute agency to all sorts of things when none is there, it has been argued, may also have something to do with the origins of religion (Guthrie 1993; Dennett 2006; Haidt 2012).

**A brief aside: Error management theory application 2—playing it safe around strangers**

Error management theory may also be applicable to the human tendency to "play it safe", a behavior frequently observed in one-shot games when people are paired with complete strangers.

The ultimatum and dictator games discussed here are basically one-shot experiments (they involve a single interaction). And as we have seen, participants in these experiments tend to behave fairly instead of pursuing maximal immediate gains. We humans seem to have a tendency to play it safe in one-shot games and interactions with strangers. David G. Rand, who studies cooperation, refers to this as social heuristics (Rand et al. 2014).[16]

A major difference between one-off interactions and those in which the same situation takes place over and over, like in repeated games (games with several rounds), lies in the number of clues we get. When a situation is repeated over and over, you gain more and more knowledge with each iteration about what happens when you make a particular choice and what the implications of that are. You're constantly thinking, when I did this, that happened, and when I did that, something else happened, and so on and so forth. But when you've just met a stranger for the first time or you're participating in a one-shot game, such clues are naturally a lot scarcer.

According to error management theory, we have to consider the error cost of each option. In the dictator game, that means the respective costs of choosing to either appear kind to others by offering the stranger some money or appearing entirely selfish by hogging the money for yourself. To determine the likely error cost of each behavior, we need to figure out what will happen, and what the implications will be, if we select that behavior. Your past experiences, the cultural, social norms and the like will play an important role here. Your decision about which option carries the lower cost will no doubt take such factors into account.

When trying to figure out which option has the lowest cost in situations where clues are scarce, such as one-shot games, you will probably try not to stray too far

[16] **Heuristics** are mental shortcuts used to simplify the way we make decisions when presented with problems. You can think of this as intuitive reasoning. There are many types of heuristics. Social heuristics are strategies that people internalize because they are typically advantageous and successful in their daily social interactions (Rand et al. 2014).

from the prevailing social norms. In other words, you will probably lean toward behaviors you consider to be relatively safe in the current cultural context. Recall that we discussed findings from ultimatum games conducted all over the world in Sect. 2.2. The way participants behaved in the one-shot scenarios was connected to the normal everyday behaviors present in each culture.

Research has also shown that people are more cooperative when put under time pressure and thus forced to make intuitive decisions; for example, they donate more (Rand et al. 2012) and make fairer proposals in the ultimatum game (Cappelletti et al. 2011).[17] It would seem that the less time someone is given to make a decision, and thus the harder it is for them to discover clues, the more they are likely to want to play it safe.[18] Researchers have also reported that people who make intuitive decisions to cooperate are perceived as more trustworthy than those who cooperate in a calculating way based on a rational assessment of the costs (Jordan et al. 2016c).

A situation in which suitable clues are scarce can also be characterized as highly uncertain because it's difficult to predict what the outcome of your actions will be. Indeed, research shows that when uncertainty is increased such that people find it difficult to predict outcomes, cooperation increases, even in situations where cooperation can never be beneficial (Van den Berg et al. 2021).

When we have few clues as to how we should act, such as when interacting with a stranger or playing a one-shot game, it is quite possible that playing it safe is still the logical course of action even if that behavior sacrifices an immediate gain and would thus seem to result in nothing but a loss, because the error cost is small (Kobayashi, mimeo).

## 5.4 What Games Have Taught Us

**Do market forces make people fair?**

To round this chapter out, I would like to go over some very interesting findings from ultimatum and dictator game experiments. These findings tell us that people are more likely to be fair to strangers in societies where they commonly buy and sell things in marketplaces (Henrich et al. 2004; Henrich et al. 2010a).

We buy most of our daily food and other necessities in market establishments such as supermarkets and other shops. A marketplace, or market, is simply a place where goods and services are bought and sold for money, and a market society is one in which you can generally do or have anything as long as you have the money to

---

[17] Note, though, that while meta-analysis (Rand 2016) supports the work of Rand and colleagues (Rand et al. 2012), others have reported being unable to reproduce the results (Tinghög et al. 2013; Bouwmeester et al. 2017).

[18] Others have reported that people tend to be more conservative when put in situations that make deliberate thought more difficult (Eidelman et al. 2012). This may also have something to do with what is going on here.

pay for it. I think pretty much everyone has a sort of vague sense that societies like this probably make people more selfish.

Yet research findings indicate that the opposite is in some sense true. Take a look at Fig. 5.14. The mean dictator game offer is plotted along the vertical axis, and a measure of the society's reliance on markets along the horizontal axis. A quick inspection of the graph indicates that the more a society buys and sells goods in marketplaces, the higher the dictator game offers. This suggests there might be some relationship between a society's reliance on markets (market integration) and the extent to which people treat strangers fairly in that society.

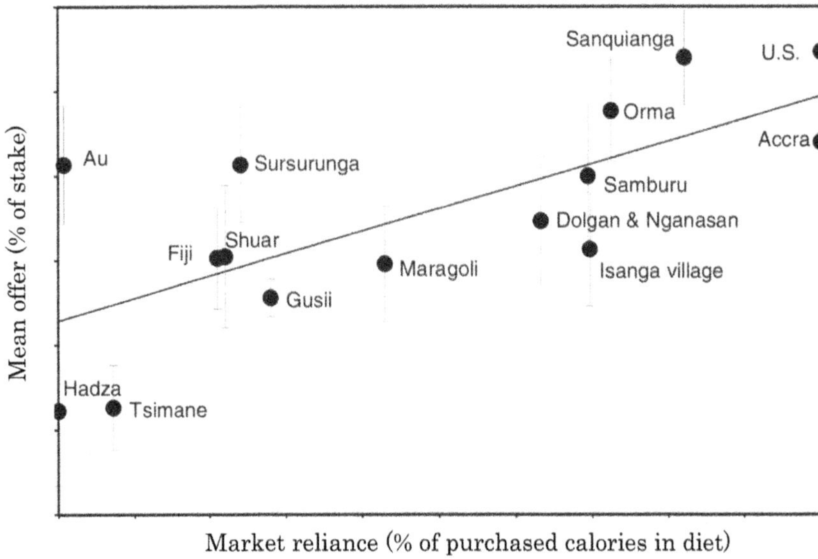


**Fig. 5.14** Mean dictator game offers plotted against market reliance. *Source* Henrich et al. 2010 (wording slightly modified)

This finding that there may be some sort of relationship between market integration and how fairly strangers are treated raised some eyebrows. The Wall Street Journal even published an article on it under the headline, "The Civilizing Effect Of Market Economics" (Wall Street Journal, January 24, 2002; https://www.wsj.com/articles/SB1011885700476777840). These were fascinating and completely unpredicted findings. Subsequent research has also shown that people living in societies that rely more heavily on markets offer higher amounts (close to half of the pot). In societies with very little market integration, people offered only about a quarter of the pot to strangers (Henrich 2020). These trends were also shown to be unaffected by income, wealth, society size, education level, and other demographic variables.

An obvious question to ask here is "Why does this happen?" Perhaps markets readily develop in societies in which strangers are treated fairly. Or perhaps, conversely, people happily exchange valuable goods and money with strangers in societies with well-developed markets, and this leads them to treat strangers fairly.

Or is it some other mechanism entirely? How do we best move closer to an answer? There is still much we don't understand and new mysteries for us to solve here.[19]

**X-raying human behavior**

According to science writer William Poundstone, the famous behavioral economist Colin Camerer has described the ultimatum game as "a cultural X-ray" (Poundstone 2010, p. 122). I'm not privy to the details of that conversation, but I believe the point Camerer was making is that just like a medical X-ray illuminates certain structures within the human body that would be invisible to us otherwise, the ultimatum game can be used as a tool to illuminate certain aspects of culture that would otherwise be quite impenetrable. Looking back on the research to date, I think it might also be fair to broaden that idea a little and call the ultimatum game, and the dictator game, which can be considered a variation of it, a *human behavior X-ray*. This simple little game has taught us so much more than we could have imagined.

**An aside about Japan's self-restraint police**

The COVID-19 pandemic descended upon us early in 2020, quickly embroiling the world in a flurry of efforts to combat the virus. Here, I present some observations about the Japanese experience with COVID-19 (or simply "Covid") that I think are relevant to the themes of this book.

To prevent the spread of this strange pathogen, people were implored to stay at home and refrain from all sorts of social activities. In Japan, these were known as "self-restraint" measures, and it is important to note that these requests for self-restraint were not legally binding. Nevertheless, anyone who, for whatever reason, went out of the house or traveled home to see the folks, and even children who simply went to the local park to play, came under intense criticism. Self-appointed "self-restraint police" would seek such people out and denounce them. Severe discrimination against anyone unlucky to become infected also became a major social problem. I'd like to examine these problems briefly within the framework of this book.

We discussed the finely tuned cheater detection capabilities of humans in Sect. 5.1, where I explained that we are very quick to spot cheaters in situations where cheating is a distinct possibility and when the cheating is both intentional and beneficial to the

[19] Subsequent work has produced both similar findings (Baldassarri 2020) and findings that are not entirely consistent with those discussed here (Cochard et al. 2021).

Other work, however, has found associations between the degree of market interactions in a society and the way prosociality manifests (Enke 2022), and a study using data on 49 countries found that market-oriented societies have a greater aversion to unethical behavior, higher levels of trust, and are not significantly associated with lower levels of morality (Callais et al. 2022). The word "dehumanization", meaning to deny someone else's humanity or not see them as human, will come up in the Chap. 6 supplement, but research suggests that interacting with strangers in a market actually makes us come to humanize other people—we use more moral and bonding language with respect to them, for instance (Harris 2023). Other studies also have strongly suggested that the market fosters a certain type of prosociality (see Henrich 2020 for details).

Finally, Jacobs (1994) examines market ethics and political ethics as two types of moral rulesets for human society, arguing that the former in particular is important to how we cooperate with strangers.

cheater. We also discussed the possibility that people are more sensitive to cheating the more it is potentially detrimental to themselves.

With a strange pathogen on the loose, anyone heedlessly out and about town might seem to meet all of these criteria. Leaving the house was legally permissible, so the opportunity to cheat existed, and anyone moving about public spaces was almost certainly doing so under their own volition, so the cheating was intentional. Above all, leaving the house for your own convenience or pleasure would clearly constitute cheating that benefits the cheater. Moreover, the risk of people spreading the virus by entering public spaces meant that this form of cheating was perceived as highly detrimental by those seeking to denounce it. In short, people were on high alert for this sort of cheater, and research has shown that people experience anger at cheaters.

We also discussed the ability of gossip to rapidly and widely expose moral and social transgressions. Stories like "That guy did this horrible thing!" potentially stimulate our brain's reward system, giving us a sense of joy. Gossip is also more pervasive when the group faces a problem it must deal with. It is a policeman and a teacher, just like Japan's self-restraint police. It polices people who do the wrong thing, and it teaches them, for example, that "leaving the house during a pandemic is not okay!".

We also discussed how people seldom forget who cheaters are. If caught, you run the risk of forever being the subject of chatter like, "That's that guy from down the street; he was so selfish during Covid!".

People feel strong anger toward and show a strong desire to punish cheaters, as discussed in Chap. 4. And that thirst for punishment is stronger when the cheating is perceived to be intentional. It should thus be evident from what we've discussed that we feel compelled to admonish cheaters, and that we have an intense desire to impose punishment on cheaters, particularly punishment that is effective (actually harms the cheater) as opposed to symbolic, even at a cost to ourselves. That is, the suffering of cheaters gives us joy and pleasure. (As will be touched on in Chap. 6, it may be the case that we dehumanize cheaters in the sense that we see them as separate from our own social group [we put them into the out-group category] and thus shut off faculties of social cognition with respect to them.)

Moreover, this strong desire to punish is not necessarily associated with a direct interest in the interaction or behavior. Even if the cheater is a stranger in some faraway place and thus wholly unlikely to affect us directly, this has very little bearing on our response; we still feel strong anger and resentment and a strong desire for punishment.

Meanwhile, few people are willing to directly punish wrongdoers, apparently for fear of retaliation. As pointed out, punishment in real life usually takes a form that is not too costly to the punishers, such as gossip. We also touched on the possibility that gossip serves as a mechanism for identifying wrongdoers, making it possible to deal with cheaters collectively. That is, one feature of gossip is that it makes it possible for us to *gang up* on norm violators so that punishers need not step forward as individuals (put simply, they can hide behind anonymity), which protects them from the threat of retaliation.

It has also been reported that when people pass along high-quality, juicy gossip, they feel more powerful and they feel more closely connected to their gossip partners.

Gossip not only seems to reduce stress hormones, but it also helps us bond with others. The prevalence of terms life "self-restraint" and "social distancing" in our consciousness and the need to stay physically separated from others no doubt left a lot of people with feelings of isolation. Gossip may have been quite valuable as a means of relieving stress and helping us feel connected with someone under these circumstances.

When gossip levels intensify in response to a problem faced by the group, the stories passed around concern not only bad actors but good ones as well, and it is argued that the purpose of this is to give encouragement to upstanding members of the group. So at the same time as it singles out wrongdoers, gossip may also function as a means for the "good" people of the group to positively affirm their own behavior—"We don't behave like that, we're all doing the right thing!".

During the pandemic, I heard a lot of people say something like, "Covid is scary, but I'm even more worried about people finding out that I got it!" This state of mind is perhaps understandable in light of the preceding discussion.

Of course, I have no doubt that this aspect of human cognition played a positive role in terms of protecting society from cheaters by curbing the spread of the virus and so forth. It has no doubt played an integral part in protecting our societies over the long process of evolution. And yet, as Dunbar says, "punishment is a poor way to create social cohesion." Excessive condemnation and discrimination is most certainly problematic.

Finding a balance between those two opposing forces would be ideal. But as should be evident from the research we've discussed, these things are deeply ingrained in us as humans, and this realization, I think, tells us that the path to solving these issues is no easy one. Nevertheless, it is my hope that knowing that these tendencies are etched into our brains will be the first small step toward a more considerate and compassionate society. As the moral philosopher Joshua Greene puts it, "Human empathy is fickle and limited, but our capacity for empathy may provide an emotional seed that, when watered by reasoning, flowers into the ideal of impartial morality" (Green 2013, p. 201).

**Summary**

Abilities etched into our brains

- We are able to quickly detect cheaters.
- We quickly spot intentional cheating that benefits the cheater in situations where cheating is a distinct possibility.
- We find it easier to spot cheating by lower-ranking individuals than cheating by higher-ranking individuals.
- The more costly the cheating is to us, the quicker we spot it.
- We quickly spot hazardous behavior regardless of whether it is intentional.
- We seldom forget the faces of and pertinent information about cheaters. We also remember the faces of people with high social status.
- The roles of gossip

1. Facilitates reputation management: helps us find out about our own reputation and the reputations of others
2. Identifies social deviants
3. Serves as a policeman and a teacher

    Policeman: reprimands bad actors through reputation
    Teacher: teaches us what behaviors are considered bad

4. Creates interpersonal bonds
5. Gives positive reinforcement to good people

- Information about cheaters quickly spreads via gossip.
- People are attuned to avoid more serious (costly) mistakes.

  What the ultimatum game and dictator game have taught us.

- The more a society relies on markets, the more it treats strangers fairly.

Our brains are equipped with an astounding ability to detect cheaters and the ability to remember their faces and other pertinent information about them. And that information about cheaters spreads rapidly via gossip. In a society like this, being acutely worried about reputational effects and the watching eyes of others, and being concerned with appearing to be good, even if this means sacrificing immediate gains, may ultimately be in our long-term interests.

## Chapter 5 Supplement

### The Wason selection task

In Sect. 5.1, we considered cheater detection and hazard management problems through the lens of the Wason selection task. Researchers have also used the Wason selection task to study other interesting ideas, including altruistic detection (Brown and Moore 2000; Oda et al. 2006).

It has also been argued that cheater detection and hazard management are not individual skills but instead both part of the human ability for general reasoning about obligations and permissions (what one must do or what one may do). Based on this idea, cheater detection and hazard management can be viewed as components of the broader problem of detecting violations in the context of obligations and permissions, rather than as separate, individual problems.

Leda Cosmides and her former graduate student Laurence Fiddick have raised various objections to this idea.

For example, when the context involved permissions but not cheater detection, the correct answer rate was only around 30–45%, which is not all that high. Similar results have been reported in Japan (Hasegawa and Hirai 2000).

Moreover, and as explained in footnote 39 in Chap. 5, cheater detection and hazard management problems evoke different emotions (Fiddick 2004), and different regions of the brain are associated with each (Stone et al. 2002; Fiddick et al. 2005). These findings suggest that the brain may perceive these two types of tasks as completely different problems (Fiddick 2011).

In light of these findings, Cosmides and colleagues argue that we humans have separate faculties for dealing with the tasks of avoiding deception by cheaters and protecting ourselves from hazardous activities, two important problems we faced in our evolution, rather than a single, unified ability that encompasses both of these faculties.

**The brain's reward system**

To round out the discussion on brain research, let's take a brief look at the brain's reward system.

The region of the brain activated when we meet our basic survival needs for water, food, and the like is the reward system, which is associated with feelings of joy and pleasure. The striatum is a critical component of this reward system. It is thought to provide the basic mechanism that motivates us to strive for and consume food and water in order to go on living.

Now, this brain region that responds to basic rewards needed for survival also responds to a whole slew of other things that are quite different in nature, many of which we've covered. The list includes secondary rewards such as money, social rewards in the form of positive reinforcement from others, the reduction of disparities, being treated fairly, the pain of a rival sports team's fans, the misfortune of those we find envious, the suffering of wrongdoers, and negative gossip about celebrities.

Researchers have also reported that the striatum responds when we give charitable donations, particularly voluntary ones (Harbaugh et al. 2007), when we cooperate with others (Rilling et al. 2002), and when a mother sees her baby smile (Strathearn et al. 2008). The list goes on and on. And, as a human, you can probably relate in some way to every single one of these things.

Hence, the brain activity associated with self-serving rewards, such as our own resources and income, and social or altruistic rewards, such as fairness, cooperation, and serving others, overlaps in the striatum. Looking at what the reward system responds to may reveal, for better or worse, what our brains are directing us to seek out (of course, as human beings with the power of reason, the question of whether we should unquestioningly follow those directives is another matter).

These brain reactions are not all identical, however. If you look closely, there are naturally all sorts of differences in terms of what areas of the striatum are activated and what other regions of the brain are activated. Plus, brain research is a field that has been developing rapidly in recent years, and there is a very real chance of past results being reinterpretted and experimental findings turning out the be non-reproducible (this is, of course, a problem common to many other areas of research, including those covered elsewhere in this book, and not limited to brain research). Experimental reproducibility refers to whether the same results can be obtained when an experiment is repeated. It is particularly relevant to experimental scientific work in general and something we need to take proper care with.

As neuroscientist Naotaka Fujii puts it, what makes social neuroscience fascinating is "not just that it can help us understand how the brain works, but the possibility that this can in turn help us to understand people and society" (Fujii 2010). It's an important field of study and one that I will continue to keep a close eye on as exciting new research developments roll out.

# Chapter 6
# The Light of Evolution

## 6.1 Adaptive Rationality

**Dissecting the 20% sliver of hope**

Once again, let's briefly recap the discussion so far.

We have been examining something called the ultimatum game, in which money is split between two players. The proposer (Player 1) is given a stake, commonly an amount like $10, and asked how he proposes to divide it with a stranger. The responder (Player 2) either accepts or rejects that proposal. If Player 2 accepts it, the money is split according to the proposal. If Player 2 rejects it, the $10 is confiscated and both players walk away empty-handed.

If both players are Econs, Player 1 will propose ($9.99, $0.01), and Player 2 will accept this proposal. The basic problem, however, is that this outcome is intuitively unconvincing, and indeed, in many experiments, Player 1 typically offers 40–50% of the stake to Player 2.

Perhaps Player 1 is afraid of Player 2 saying no, you might think. But even in dictator game experiments, which remove Player 2's right to accept or reject the proposal, Player 1 typically gives around 20–30% to Player 2. This most curious result has been called a "20% sliver of hope" for humanity, but what constitutes that 20% sliver hope, what are its inner workings? That's the question I've been tackling most directly in this book.

**Humans evolved to be members of society**

Ultimately, the answer (or part of it at least) is that humans cannot survive alone. Humans could only survive and thrive in groups and thus developed the ability to empathize in order to maintain cooperative relationships within their groups. Empathy alone is insufficient for dealing with free riders, however, so the human brain was wired to feel anger at unfairness and joy in punishment, along with the ability to detect and remember cheaters.



K. Kobayashi, *The Mystery of the Ultimatum Game*,
https://doi.org/10.1007/978-981-97-8979-5_6

Once you're labeled a cheater, your pain and suffering are quite agreeable to others around you. They are happy to punish you even if they have no direct vested interest in the interaction or behavior that involved cheating. What's more, people quickly spot cheating and quickly spread relevant information about it throughout the group in the form of gossip, and the label of cheater sticks with you (is not forgotten) for a long time.

I need not tell you just how risky behavior that might be perceived as cheating is in a society like this. Humans thus became very sensitive to eyes as a signal of being watched. Humans, who derive joy from positive connections and who fear isolation as if it were real pain, evolved to live as members of a society.[1]

**Adaptive rationality**

Put simply, these are all survival strategies, shaped by the ancient environs in which our evolution took place for the purpose of enabling us to survive and reproduce (Fig. 6.1).

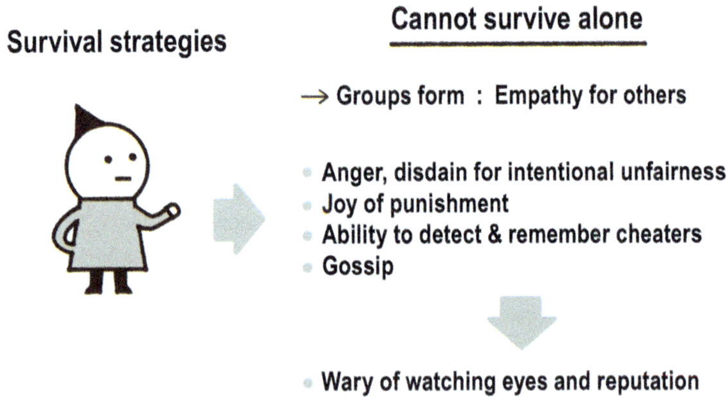


**Fig. 6.1** Survival strategies[2]

Recall that at the beginning of this book, I said that humans are not rational in the way that traditional economics assumes, but that they are not, as is often claimed in recent years, irrational either; they are, in fact, *very very rational*. You may have

[1] Note that social group here refers to the one to which you belong—your in-group—and not groups to which you don't—out-groups. See the end-of-chapter supplement for more information.

[2] I think there are other important elements to the inner workings of the 20% sliver of hope aside from what we've discussed in this book, namely that our desire not to be ostracized from our group makes us wary of the watching eyes of others.

For example, regardless of whether someone else is there to see it or not, and regardless of whether there is any possibility of other people knowing about it, don't you feel unsettled at the idea of behaving as if you only ever think of yourself? Don't you find the idea of being utterly selfish entirely off-putting and intolerable? I call this the internalization of norms, and I think this is also a very important part of the workings of the 20% sliver of hope. I have, unfortunately, omitted any discussion of this from this book out of space considerations. I would love to put all of my thoughts on this together at some point, though.

wondered at the time what on Earth I could possibly mean by "very very rational". If you've made it this far in the book, though, perhaps my meaning is starting to fall into place.

To be sure, humans do on occasion behave in ways that may not seem rational at first blush. The ultimatum and dictator games are great examples of such behavior. If you think purely about the immediate gain, you can immediately see that there are more beneficial options available to you, yet many players choose to "take a loss" anyway. These behaviors, while not seemingly rational at first, can be regarded as highly rational in a different sense—when seen as part of strategies for surviving and reproducing. My aim throughout this book has been to demonstrate that.[3]

In this book, I would like to propose a concept called **adaptive rationality**, meaning rational in terms of adaptation. This is the idea that we tend to follow behavioral patterns that increase the likelihood that we will survive and reproduce (in the context of our ancestral environment), instead of acting in a way so as to directly acquire resources as traditionally assumed in economics. Adaptive rationality is what I meant when I initially said "exceedingly rational". One might even call this irrationally rational[4]! We humans certainly do behave in predictably irrational ways at times. But this might just be because we are very very rational.

**Deep rationality**

This idea of adaptive rationality that I put forward in this book rests on essentially the same reasoning as the concept of **deep rationality** advanced by evolutionary psychologist Douglas T. Kenrick and colleagues (Kenrick et al. 2009; Kenrick 2011; Kenrick and Griskevicius 2013).

Kenrick makes two main claims. The first is that human decisions serve evolutionary goals. The second is that human decision-making is designed to achieve several completely different evolutionary goals. He identifies seven crucial evolutionary challenges in this regard: (1) evading physical harm, (2) avoiding disease, (3) making friends, (4) gaining status, (5) attracting a mate, (6) keeping that mate, and (7) caring for family.

So the humans who left more offspring and thus became our ancestors were "those who protected themselves from enemies and predators, avoided infection and disease, got along with the other people in their tribe, and gained the respect of their fellow tribe members. They also successfully attracted a mate, established a partnership with that person... [and] cared for their needy and relatively helpless offspring"

---

[3] It has also been shown that such behavior is rational in the sense that it can be considered to be the result of consistent, properly reasoned choices that take the interests of others into account, and not simply the result of errors or random choices. See the end-of-chapter supplement for more information.

[4] Many authors have discussed the possibility of seemingly irrational behaviors actually being rational in some specific sense. See, for instance, Frank (1998), Gigerenzer (2007), Haselton et al. (2009), Kenrick et al. (2009), and Kenrick and Griskevicius (2013). The economist Robert H. Frank posits that the role of the emotions is to guide our behavior in a more rational direction beyond immediate rewards or immediate gratification. These perspectives, of course, accord in some respects with those I present in this book.

(Kenrick and Griskevicius 2013, p. 30).[5] These sorts of evolutionary challenges are also sometimes called **adaptive challenges.**[6] It is extremely important to consider human rationality from the perspective of adaptive challenges.

While there may still be considerable discussion about whether these seven items comprehensively cover all crucial adaptation challenges, there's little doubt that those humans who were less successful at solving these challenges failed to become anyone's ancestors. To survive and reproduce, our ancestors would have had to constantly solve these challenges. And on that basis, one should naturally expect the cognitive mechanisms for confronting these challenges to have been engraved into the human mind over the vast stretches of time during which our evolution took place. Evolution has equipped our minds with an array of mechanisms for dealing with a myriad of challenges.

There are some crucial points to note here. We no doubt faced multiple adaptive challenges, and they each may have given rise to different behavioral patterns that seem inconsistent or contradictory. Yet behind these complex and seemingly inconsistent behavioral patterns are evolved adaptations that increase the chances of us surviving and reproducing, and we can thus take a unified view of these behavioral patterns.

**Rationally irrational, rationally rational**

Let me be clear, however, that the concept of adaptive rationality discussed in this book is a broader concept in the sense that it is not necessarily inconsistent with the conventional economic notion of rationality.

In this book, adaptive rationality means being rational in the sense of increasing the chances of survival and reproduction (in the context of our ancient environments). Naturally, obtaining more of a valuable resource like food or money is highly appealing from both a survival and from a reproduction standpoint. If that's the case, then if decision-makers are adaptively rational, then they should definitely be capable of behaving in a manner that is "rational" given the situation at hand.

Recall, for example, the discussion of anonymity in Chapter 3. We looked at how eliminating the sense of being watched to an absolute minimum resulted in people choosing rational behavior reminiscent of an Econ. When people felt it was possible to obtain a valuable resource right in front of them, that's actually what they did. So it must be that we humans also have a rationally rational aspect to our decisions and behavior too.

Rationality in an adaptive sense should tend to bring about traits that are adaptive over the long run. If such traits are preferable in the long run, they may result in a tendency to behave in ways that initially seem to produce nothing but a loss. If,

---

[5] The technical term for this is "fitness", the precise definition of which is a bit involved. In very simple terms, those with a strong ability to survive and reproduce are said to be highly fit. See the end-of-chapter supplement for a more detailed explanation.

[6] The cognitive mechanisms adapted to solving these challenges are called **modules**. See the end-of-chapter supplement for a brief explanation.

however, they do not cause any real problems in the long run, then they should produce a tendency to choose these sorts of simply beneficial behaviors.

A decision-maker who is adaptively rational can behave both in ways that are rational and in ways that initially seem to be irrational. So not only is the concept of adaptive rationality I present here not incongruous with behaviors that may initially seem irrational, but it is also not in conflict with Econ-like rational behavior. Indeed, it is crucial to note that it encompasses both.

Behaviors that seem rational and behaviors that do not seem rational have, until now, naturally been treated as the inverse of each other, as incongruous with each other. But in light of the above, we need not see them that way. We humans can, depending on the situation, be both rationally irrational and rationally rational. Hence, the concept of adaptive rationality advanced in this book is a broad one and encompasses both rational irrationality and rational rationality, two aspects that may at first seem incongruous with each other.

This sort of breadth is extremely important in the contemporary study of economics. As I have mentioned, traditional economics is based on the concept of rationality. Behavioral economics, which is very much in vogue today, is based on the idea that humans possess bounded rationality and arose as a critique of traditional economics.

The two would seem to be incompatible at first glance, but they can be unified through the lens of adaptive rationality. The concept of adaptive rationality advanced in this book provides, for the first time, a theoretical framework for a consistent, unified treatment of both traditional economics and behavioral economics (Fig. 6.2).

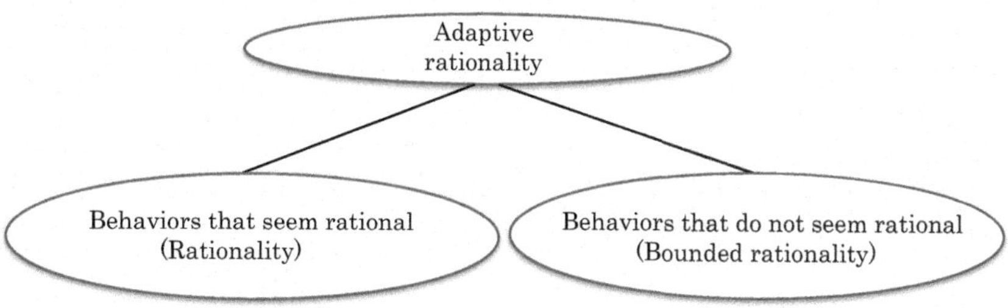


**Fig. 6.2** Adaptive rationality

## Humans as a social and a selfish species

I've often heard the idea of the Econ referred to as a benchmark model in recent years. It's seen as a standard of sorts.

As a benchmark model, the concept of the Econ most certainly is highly significant. Take the ultimatum game, for instance. The existence of the model of rational Econ-like decision-making is what prompted us to think about why human behavior deviates from the "rational" way an Econ would behave. It is what provided the starting point for this discussion.

Yet I don't think that's the only role that the Econ model has to play. There exist many cases in which traditional economic models have been able to explain the behavior of humans in society. Recall that in the absence of watching eyes, for instance, the behavior of people in experiments was quite close to what an Econ would do. This is perhaps one reason why economics has continued to use the Econ model for such a long time.

So are we Econs or are we not? Well, wouldn't you agree that viewing it in such a binary choice seem to be no longer relevant? Given the results of the myriad of ecperiments conducted so far, it seems safe enough to say that one person would behave like an Econ in certain situations and not in others. In other words, any one individual may behave like an Econ or not like an Econ depending on the situation. To behave in seemingly inconsistent ways is to be human. As De Waal puts it, "There is both a social and a selfish side to our species" (De Waal 2009).

The notion of seemingly irrational behavior is not a monolithic one either. Within the space of seemingly irrational behavior exist behaviors that would appear to be inconsistent with each other. All of our complex behavior patterns—from seemingly rational to seemingly irrational behaviors, and on to seemingly mutually inconsistent behaviors—can be viewed from the single, unifying perspective of what is rational from the perspective of adaptation. That's the major claim of this book. In that light, the Econ is not dead, nor is there any need for it to die. Rather, it continues to live on in a slightly different form.

One of the next questions to address is that of where the boundary conditions lie. When do humans behave like Econs, and when do they behave unlike Econs? Where do the boundaries of each type of behavior lie? The case of watching eyes examined in this book is but one example. There is plenty of other research seeking to probe the boundary conditions by showing how humans behave differently depending on the situation at hand.

The research to date has, in a way, tended to focus solely on anomalies. I am certainly not denying the importance of anomalies, but at the same time, I think trying to figure out just when it is that humans display rationality of the traditional economics variety is another crucial avenue of inquiry. Figuring this out would bring the boundary conditions into much better focus. To advance our understanding in this regard, we need to probe the underlying causes of these behaviors. That is, we need to ask why humans have the innate behavioral tendencies that they do. This is a matter of exploring what are known as the ultimate factors. I believe one of the keys to this problem lies in the adaptive challenges that we humans have faced over the course of our evolution, as briefly mentioned in this book.[7]

---

[7] To summarize the discussion so far, there appear to be several reasons for seemingly irrational behaviors.

There are several possible reasons just within the scope of this book, including small errors that inevitably result from efforts to prevent large errors in accord with error management theory, behaviors that were adaptive in the past but are no longer so, behaviors resulting from the contradictory workings of multiple modules, spandrels (the byproducts of adaptations), and outcomes that are simply the result of chance.

### Modern skulls house stone age minds

I'd like to round out this section with a few words on the term "adaptation". In recent years, we've come to see our brains and minds as products of evolution, just like our body. The mind evolves, in other words. In biology, **adaptation** is the process by which organisms, through natural selection, acquire advantageous biological traits that make them more likely to survive and reproduce in a given environment.

The stage on which human evolution took place is basically thought to have been one of hunter-gatherer lifestyles.[8] Hence, the idea of adaptive rationality is that the rationality we humans possess was etched into our minds over the long process of our evolution in an ancient environment. As the pioneers of evolutionary psychology Leda Cosmides and John Tooby put it, "**modern skulls house stone age minds**" (Cosmides and Tooby 1997).

It is crucial to remember that the backdrop to this process of adaptation was basically an ancient environment. Our brains, bodies, and minds were shaped by and to be suitable (adapted) for life in the distant past, back when our evolution took place. Turning this around, so to speak, adaptively rational decision-making agents well-adapted to life in an ancient environment may, simply by acting in ways that evolution has shaped them to act, frequently choose behaviors that turn out to be maladaptive in our modern environment. One example of the result of such a behavior that is often trotted out is obesity. See the brief aside below for a discussion.

---

When we give explanations of the direct mechanisms of human behavior like "Humans worry about eyes and reputations, so they react to watching eyes", we are explaining behavior in terms of **proximate factors**. On the other hand, when we ask why humans have a mechanism for worrying about eyes and reputations in the first place, we are trying to get at the **ultimate factors**.

I guess one might describe the research literature at present as a swirling sea of anomalies. Nevertheless, one cannot help but acknowledge that there is very little research tracing back to the ultimate factors behind them, and such perspectives are expected to become increasingly important ahead. At the same time, we need more detailed discussion on the mechanisms by which these anomalies occur, and on organizing or categorizing these anomalies according to some standard to facilitate a unified analytical approach.

The hundreds of anomalies out there include many that are seemingly contradictory at first glance. But I believe we can work toward a unified explanation if we approach these anomalies from the perspective of adaptive rationality.

[8] The technical term for the stage on which human evolution took place is the **environment of evolutionary adaptation** (EEA).

Our EEA is basically thought to have been that of ancient hunter-gatherers, and the term is often used to refer to the Pleistocene era, the span of several million years from the advent of the genus Homo, which led to modern humans (Homo sapiens), up until the appearance of agriculture. It depends, however, on the type of adaptation being considered, so the term does not necessarily refer to a specific time or place (Buss 2019).

**Fig. 6.3** Modern skulls house stone age minds

**A brief aside on obesity in an evolutionary context**

Obesity is an oft-cited example of the outcome of a behavior that was adaptive in the past but is no longer adaptive in our modern environment (Power and Schulkin 2009).

With our modern lifestyles, if you were to eat as much as you wanted of whatever you wanted every day, you could very quickly develop obesity and other lifestyle-related diseases. Surely we would not do this if our bodies simply had a mechanism for making us feel full when we'd eaten just the right amount. The reason why there is such a disconnect here when it comes to something so fundamental and directly connected to our survival as eating has to do with the ancient environment in which our ancestors lived.

Humans in the ancient world had no idea when their next meal would be available. If you encountered a valuable source of calories like sugary or fatty foods, you'd eat a whole lot of it simply because it tasted good, and then you'd eat some more if any were left. Storing the excess calories in the form of subcutaneous or visceral fat would have been advantageous for survival. Individuals who left valuable food behind because they felt too full were surely putting themselves at a disadvantage. In this sense, the trait of being able to eat more and more without feeling full even after having already had "enough" is thought to have been adaptive in our ancient environment.

Yet our bodies apparently don't have a mechanism for monitoring the build-up of excess nutrition and letting us know that we've had enough. This is probably because we never encountered circumstances in our ancient environment in which we had as much food as we could possibly want available so often and so readily enough as to make a regulating mechanism like this advantageous. Levels of physical exercise are also vastly different in our modern world. In our ancient environment, acquiring food would have required quite a lot of physical effort, but today, taking a trip to the supermarket is probably the most strenuous thing we do in pursuit of food, and you can even get by just by twiddling with your smartphone. So while a mechanism that encouraged us to gobble up sugary and fatty foods and store the calories in our bodies may have been adaptive in our ancient environment, in our modern world, not only is it not adaptive, it may even be potentially harmful.

These sorts of diseases linked to adaptations that were formed in our ancient environment but have deleterious effects in a modern setting are now widely referred to as mismatch diseases (Lieberman 2013a). There are many more candidates for the status of mismatch disease than you might think. They include type 2 diabetes, tooth decay, Alzheimer's disease, cirrhosis of the liver, asthma, athlete's foot, hypertension, gastric ulcers, and some cancers.

## 6.2 The Evolution of Cooperative Behavior

### True altruism

In this book, we have examined a prime example of an anomaly in game theory: the altruistic behavior of Player 1, the proposer, alongside the proposal-rejection behavior of Player 2, the responder, in the ultimatum game. We have looked, in particular, at Player 1's seemingly altruistic behavior of sharing something of value with a stranger even at a cost to himself, something I've referred to as the "20% sliver of hope" for humanity. And probing the workings of this 20% sliver of hope has been a direct theme of this book.

The conclusion here is that this is the result of Player 2's proposal-rejection behavior motivated by intense disgust and anger at being treated unfairly and, against that backdrop, the behavior of Player 1 reflecting an intense preoccupation with watching eyes and reputational effects. In short, it's a sort of **superficial altruism** on the part of Player 1, who really doesn't want to be seen as stingy and selfish.

Let me add one point here. This is completely different from the claim that everyone is a hypocrite underneath. All I am saying is that in abstract, neutral situations like the ultimatum and dictator games, behavior that can only be described as superficial altruism plays enough of a part that we cannot ignore it.

We can all, no doubt, remember having felt a desire to help someone in need, even a complete and utter stranger. We humans feel compelled by this mysterious and yet intense feeling to do something for someone else at times. That we have this tendency is, itself, something about which there is little doubt. Let's call altruism driven by this feeling "true altruism". The behaviors that flow from true altruism and superficial altruism may often appear identical at a surface level. And rigorously delineating between the two may be no easy task. Yet this does not obviate the need to draw a distinction between them when discussing our behaviors, and they are certainly not mutually exclusive. Just because we humans have a capacity for superficial altruism that merely makes us appear to be good externally does not mean that we do not also have a capacity for true altruism driven by that mysterious feeling described above.

In the words of famous behavioral economist Ernst Fair, "Our species is apparently the only one with a genetic makeup that promotes selflessness and true altruistic behavior" (Fehr and Renninger 2004). While I do feel a little hesitation at the phrase "the only one", this is the view taken in this book as well. And there is no end to the puzzles here. Just what is true altruism anyway? When does it show up? Are we really the only such species?[9]

### The paradox of altruism—red in tooth and claw

Finally, I'd like to give a brief background on the themes of this book.

In this book, we have examined the cooperative nature of humans in the form of altruism and morality through the lens of a simple task: how to divide an endowment of $10. "A key aspect of human behaviour is cooperation," note renowned mathematical biologist Martin A. Nowak and colleagues (Dreber et al. 2008). And the renowned economists Samuel Bowles and Herbert Gintis, noting that we are "an exceptionally cooperative species" (Bowles and Gintis 2013), observe that the salient characteristics of human cooperation are the scale on which it occurs, the complexity it involves, and the fact that we extend it to complete strangers.

You may have heard of Leonard E. Read's famous essay "I, Pencil", the story of how an ordinary wooden pencil is made, told in the words of the pencil itself (Read 1958). Nobel laureate and famous economist Milton Friedman describes the tale fondly in the book *Free to Choose* (Friedman and Friedman 1990). The pencil starts out with the fantastic statement that "not a single person... knows how to make me" and then proceeds to tell a delightful tale of the complicated cooperative relationships through which wood from a tree is turned into a pencil that ends up in a person's hand. This process involves untold thousands of people from different countries and ethnic backgrounds—people to cut down the tree, to mill the wood, to obtain the material for the pencil core (the "lead"), to transport the materials, to make the tools for doing all this, and so on and so forth. This complicated process of furnishing something as simple as a single pencil embodies human cooperation on a grand scale. This, for me at least, brings into stark relief just how much our everyday lives

[9] The initial motivation for this book were the looks of incredulity on the faces of my students when confronted with the ultimatum game. And those looks were not so much prompted by the discrepancy posed by Player 1's seemingly altruistic behavior as by the discrepancy posed by Player 2's willingness to reject proposals at a personal cost.

As we have discussed, one could, it seems, explain Player 2's actions by saying that she rejected the proposal because she felt intense anger and disgust at the other player's selfishness. This sort of explanation in terms of the direct mechanism is an explanation based on the proximate factors. An explanation based on the ultimate factors, meanwhile, would tell us why we have a mechanism for feeling intense anger and disgust at selfish people in the first place.

What this means, therefore, is that in this book we have mainly discussed the ultimate factors for Player 1's behavior (and in the sense that we have been trying to probe the workings of the 20% sliver of hope, we have been focused on the ultimate factors behind Player 1's behavior in the dictator game, which, again, is a simplified version of the ultimatum game). When it comes to Player 2's behavior, however, although we may have been able to explain the proximate factors to some extent, this book has offered very little discussion of the ultimate factors (for more discussion, see Kobayashi 2021). My exploration of the ultimatum game puzzle has still only just begun.

during the COVID-19 pandemic, something we may have taken for granted, hinged on the cooperation of essential workers and countless other faceless individuals.[10]

That's not all. Major natural disasters, for instance, prompt large amounts of donations from people living far away, and many volunteers rush to the aid of people they've never met. And we've all seen countless news stories of people risking their lives to help complete strangers. These sorts of cooperative relationships transcending blood ties are a prominent human trait.

This is actually quite a strange phenomenon from the perspective of evolution by natural selection. This is because you would think that in the face of merciless competition for survival, individuals who act solely in their own self-interest are more likely to survive and reproduce than individuals willing to make sacrifices for strangers. This problem, known as the **paradox of altruism**, has long puzzled biologists, and it even troubled Charles Darwin, who proposed the theory of evolution by natural selection. Darwin is thought to have been the first person to discuss issues of human conscience and morality from an evolutionary perspective in direct opposition to the received wisdom of his day.

**The building blocks of morality**

Our cooperative nature has long been seen as what separates us from other animals and as an important quality that makes us human. Recent years have seen widespread debate about whether cooperativeness is an exclusively human trait. All sorts of cooperative behaviors have been uncovered throughout the animal kingdom, from bacteria to insects to chimpanzees (Ohtsuki 2014; Raihani 2021).

In the microscopic world of bacteria, for instance, multiple individuals cooperate by secreting a sticky substance to build up a biofilm that enables them to collectively stay in the location of a food source (Ohtsuki 2014). Researchers have also reported that African grey parrots engage in selfless acts to help other members of their own species (Brucks and Von Bayern 2020). Examples of self-sacrificing behavior that involves doing work or taking on risk for others also abound. A species of

---

[10] The COVID-19 pandemic that descended upon the world in early 2020 made us acutely aware of just how much we depend on countless thousands to maintain the everyday lives we take for granted. For many, the pandemic may have been their first encounter with the term "essential workers", which refers to people engaged in jobs that are essential to keeping society running.

We were all urged to stay at home to keep the virus from spreading. But what would happen if absolutely *everyone* really did stay at home? Gas, water, and electricity would be cut off. Supermarkets, butchers, and greengrocers, even your local corner store, would be shuttered. No one would be available to deliver items to your door, and no one would even be around in the first place to take your order. No one would come to collect your garbage. Hospitals and pharmacies would be closed, and not even the police or fire department would attend when needed. At some point during this part of the pandemic, the gas stove at our place broke. We were told that factory shutdowns in China meant there were no parts available, so it might not be possible to repair it, nor even to buy a new one for that matter. The scope of human cooperation extends far beyond any single country's borders.

The list of people we all depend on is seemingly endless, and eliminating the work of any single one of those individuals from the picture can threaten our blissfully normal way of life and even jeopardize our ability to protect human lives. The pandemic really brought this home for me. Our everyday existence depends vitally on the cooperation of countless strangers.

soil-dwelling amoeba called *Dictyostelium discoideum* (slime mold), for instance, are known to aggregate when their food supply becomes scarce, sacrificing around 20% of individuals so that the remaining 80% have a way to survive, and worker ants sacrifice their own chance at reproduction to cooperate with the queen ant's egg-laying and child-raising efforts (Ohtsuki 2014). Vampire bats are known to regurgitate and share blood (a valuable food source) with roost-mates that fail to feed on a given night (Carter and Wilkinson 2013). Closer to our own species, we find many examples of cooperative behaviors in primates too. Wild chimpanzees, for instance, adopt orphaned youngsters within their group (Boesch et al. 2010), and help one another against leopard attacks (De Waal 2009).

So, we know that there are "many examples of animals other than humans risking danger, expending effort, giving up food, and sometimes even giving up their lives, to benefit others" (Batson 2011, p. 117). "Our propensity to cooperate has old evolutionary roots," observes De Waal (2014, p. 53), and "the building blocks of morality clearly predate humanity" (De Waal 2005). We are now faced with the possibility that cooperativeness—something long held to be what sets us apart from other animals, the quintessential trait that makes us human—has its origins in the distant past long before humans evolved.

**Cooperation as a driver of evolution**

In 2005, Science magazine identified the question of how cooperative behavior evolved as one of the 25 most important questions facing modern science (Kennedy and Norman 2005). With the understanding that "cooperation has operated... to shape the evolution of life on earth" (Nowak 2012), it is now thought that it is our ability to cooperate in large social groups that has been one of the keys to our success as a species.

It's truly thrilling to see researchers from across the spectrum of science and the humanities come together to tackle the momentous puzzle of why we humans are such a cooperative bunch. This work transcends psychology, biology, archaeology, anthropology, animal behavior, neuroscience, cognitive science, ethics, law, game theory, behavioral economics, and more. In discussing one aspect of human behavior—that we worry about reputations and choose behaviors likely to be perceived as cooperative—my hope is that this book adds one small morsel to the mounting pile of magnificent research on this topic. It feels like a new overarching scientific discipline that seeks to illuminate just what sort of creatures we humans really are is beginning to take shape (Gintis 2009).

The more we know, the more we realize just how much we don't know. And I am so excited to find out more! I have endeavored to cover the relevant discussions in as broad a range of disciplines as I could in this book. The more we dig, the deeper and deeper the lode of puzzles seems to run, and I hope I've given you a taste of the great energy and enthusiasm pulsing through this domain of scientific inquiry. Truth be told, that has been one of the major hidden objectives of this book.

### Creating social mechanisms for real people, guided by the light of evolution

It was at the age of 20 that, half by chance, I took up the study of economics. At the time, I never dreamed I would get involved with topics like evolution, the brain, and adaptations, which, however you look at it, would not seem to have any connection to economics.

Humans are certainly not rational, but that does not mean they are simply irrational. When illuminated by the light of evolution (Dobzhansky 1973), we can see the roots of human behavior much more clearly, and the picture of humans as adaptively rational, as I claim in this book, is revealed to us.

"Economics is about being completely and utterly focused on figuring out what we can do to make people happy!" To this day these stirring words, uttered in a university lecture by a great teacher of mine, echo around my head. Economics is driven by a desire to determine what sort of social mechanisms a society's people need if they are to live happier lives. But economics is not alone here. Surely this is something that the social sciences in general seek to address.

If we are to think about what sort of social mechanisms will make people happy, we must build a real understanding of people as creatures in the real world. If we are to be completely and utterly focused on figuring out what will make people happy, then surely we must take seriously the issue of what real people are like in the real world, and we need to create social mechanisms that make sense for those real people. This is one thing that I personally feel has so far been lacking in economics.

This will, however, require cooperation among a wide and diverse range of academic fields well beyond economics and the social sciences in general. In other words, I have come to believe that the social sciences question of what social mechanisms will make people happy is something that the social sciences on their own are grossly ill-equipped to wrestle with. It matters not what we call the nascent discipline that is taking up this mantle. Call it what you will—economics, psychology, human behavioral science, or something else. What's important is that we design systems that make sense for real people in the real world.

Perhaps the idea that we should think about what sort of society will make people happy is too restrictive, though. I believe we are now in an era that demands that we all think seriously about how *all of us collectively* can be happy. That means exploring the puzzles of human behavior in the real world and figuring out what a society in which we can all live happily looks like. The social sciences must collaborate with other academic disciplines to forge this path. While I am but one researcher on my own little orbit through academia, I hope to do whatever I can to light the way, and I believe the future of our children depends on it.

## Summary

- This book proposes the concept of "adaptive rationality", which reflects rationality in the sense of increasing the likelihood of surviving and reproducing (in the context of our historical environment).

- The concept of adaptive rationality encompasses both behaviors that seem rational and those that don't seem rational.

> *The foundation of political economy, and, in general, of every social science, is evidently psychology. A day may come when we shall be able to deduce the laws of social science from the principles of psychology.*
> — *Vilfredo Pareto, 1906, Manual of Political Economy*

## Chapter 6 Supplement

### Our tendency to treat members of our own group favorably (in-group bias)

In Sect. 6.1, I mentioned that we humans evolved to live as members of social groups. Unfortunately, however, the social group to which each of us belongs does not encompass all extant humans.

As moral philosopher Joshua Greene puts it, "humans were designed for cooperation, but only with some people" (Greene 2013, p. 23). The "certain people" that Greene is referring to are the members of our own in-group, the particular group to which we belong.

Our tendency to treat members of our own group favorably or preferentially is known as **in-group bias** or **in-group favoritism**. In dictator game experiments, for instance, participants allocate more to members of their own in-group (see, for example, Yamagishi and Mifune 2008). Studies have repeatedly shown that this sort of in-group favoritism occurs even when the distinction between "us" and "them" is so trivial as to be virtually meaningless—it occurs when groups are based on supposed preference for paintings or when people are assigned randomly to groups, for example.

One study has shown that our brain response to out-group members suggests that we think more of them as objects than as people (Mitchell et al. 2006). Other studies indicate that our brain response to seeing "extreme out-group" members, such as homeless people or drug addicts, is similar to when we see something disgusting like an overflowing toilet, suggesting that we treat them as disgusting objects rather than as people. This tendency to deny the humanity of some people, to treat them as less than human, is called dehumanization (Harris and Fiske 2006; Kteily and Landry 2022). As Bloom puts it, "We shut off our social understanding when dealing with certain people" (Bloom 2016, p. 69). (It is worth noting that the tendency to treat others as objects might also arise when we simply perceive there to be a shortage of time (Jiang et al. 2024)).

I've already mentioned research showing that not only do we not feel empathy for those we see as distant from ourselves, such as a rival sports team's fans, our brain's reward system responds to seeing them in pain (Hein et al. 2010). Findings like this suggesting that we do not see out-group members as people and even find joy in their suffering reveal just how deep-rooted the problems facing our society are.

"Dehumanization remains surprisingly pervasive," observe Kteily and Landry (2022). Research continues to show us that this tendency to care only about our own group is more deeply embedded than we might have thought. This is a grave problem that also relates to issues like bullying, discrimination, and nationalism. For an in-depth discussion of these problems, see, for example, Joshua Greene's book *Moral Tribes: Emotion, Reason, and the Gap Between Us and Them* (Greene 2013), and Robert M. Sapolsky' book *Behave: The biology of humans at our best and worst* (Sapolsky 2018).

## The rationality of seemingly irrational behavior

Several studies (Andreoni and Miller 2002; Fisman et al. 2007; Andreoni 2007) have shown that behaviors that seem irrational, such as giving a complete stranger some of your resources, can actually be seen as rational in the sense that they are the result of consistent, properly reasoned choices that take the interests of others into account, and not simply the result of errors or random choices (in technical parlance, such behaviors satisfy the **generalized axiom of revealed preference**, or GARP). More recently, similar findings have been reported in Japan (Xue and Ohtsubo 2020).

However, we still do not yet fully understand what motivates us to behave in ways that take the interests of others into account. This issue of motives has gained attention in recent years (for a review, see Vesterlund 2016), with a number of potential factors being raised. Aside from the issue of **fairness** discussed in this book, these include **pure altruism**, which posits that we simply consider the satisfaction of others for its own sake, **reciprocity** (you scratch my back and I'll scratch yours), and the **warm glow** that we experience when we are kind to others.

This sort of explanation in terms of the direct mechanism is an explanation based on the proximate factors. An explanation based on the ultimate factors, meanwhile, would tell us why we have a mechanism for caring about fairness in the first place. The claim I make in this book is that we may be able to explain this (unconscious) mechanism of behavior in terms of its ultimate factors (underlying the proximate factors) by noting that such behavior increases the likelihood that we will survive and reproduce in an evolutionary context.

## On the term "adaptive rationality"

The term "adaptation" is itself prone to being misunderstood, so I would like to provide a brief supplemental explanation on adaptive rationality.

Rational from the perspective of adaptation does not mean optimal, whether in our modern environment or a historical one. Evolved adaptations tend to be fairly well put together, but there is no guarantee that they will be optimal. If something is adaptive, this does not mean it is uniquely so, either. Moreover, as touched on in this book, environments change over the vast time scales on which evolution takes place, and so traits that were adaptive in the past will not necessarily be so in a modern environment.

Further, rational from the perspective of adaptation means that something makes good sense in terms of facilitating survival and reproduction. It has nothing to do with value judgments about what is, for example, superior or inferior, or the way something ought to be.

It also does not imply that all of the workings of the human mind and body are rational from the perspective of adaptation. This is because evolved adaptations also come with byproducts and can have effects that arise simply by chance. The famous evolutionary biologist Stephen J. Gould dubbed these byproducts "**spandrels**" after an architectural term meaning the tapering triangular spaces formed by, and as a byproduct of, the intersection of two rounded arches at right angles (Gould and Lewontin 1979).

**On the term "fitness"**

It's not easy to give a rigorous definition of "fitness". Roughly speaking, it is an indicator of how well an organism is doing in a certain environment, where "doing well" means being adept at passing on your genes.

A slightly more precise definition in evolutionary biology is the number of offspring an individual will have in its lifetime, which is (very roughly) the product of an individual's probability of survival and probability of reproduction.

To put it simply, individuals that survive well and leave many offspring to inherit their genes are said to have a high level of fitness. Individuals that die early, or that survive but fail to pass on their genes, are said to be low on the fitness scale.

By implication, this means that in an evolutionary context, the genes of highly fit individuals are likely to spread throughout the population over time. That is, the kind of specimen likely to increase in number over time is said to be highly adaptive.

While it is unfortunate that these concepts are often misunderstood, please note that modern evolutionary psychology makes claims that differ entirely from those of genetic determinism, which holds that the workings of the human mind are determined entirely by genes (from birth). This book is, of course, aligned with the views of evolutionary psychology in that regard.

I should also point out that this is not the same as the idea that organisms evolve to preserve the species. The discussion around this sort of thing can be quite technical and involved, so for an in-depth look, I direct the interested reader to books like *Evolution for Everyone: How Darwin's Theory Can Change the Way We Think About Our Lives* (Wilson 2007).

**Modules of the mind**

In Sect. 6.1, we discussed the possibility that humans evolved multiple mental mechanisms or systems for dealing with challenges in our environment. These mental mechanisms or systems that attempt to deal with a particular challenge are generally called **modules**. The idea is that these modules are distinct structures within the brain that perform particular functions.

Through the process of evolution, this argument goes, our minds were equipped with a large number of such modules. But, as is often pointed out, this means that we are equipped with many different modules all of which may choose differing behaviors, such that our overall behavior seems to lack consistency at times (Kurzban 2011).

To attract a potential partner, for example, it may be effective to adopt a fun persona, to be a bit of a smooth talker, so to speak. But to hold onto that relationship once formed—to be a good long-term partner—you may need to behave a little differently. Human beings are creatures that inherently embed seemingly contradictory patterns of behavior within the same person. That's what it is to be human. There must be a beautiful side, and inevitably, an ugly side as well. The more you learn about humans, the more you realize how deep and complex and, I believe, lovable we truly are.

## The paradox of altruism and the evolution of cooperation

Self-sacrificing cooperative behavior (altruistic behavior in behavioral ecology terms) is behavior that benefits other individuals' survival or reproductive prospects even at the expense of your own. Because this jeopardizes our own chances of surviving and reproducing, you would think that such behavior would have a hard time evolving, and this is what lies at the root of the altruism paradox.

Here, we take a brief look at three main ideas about the evolution of cooperation.

The first idea is **kin selection** (Hamilton 1964). This seeks to explain cooperation between genetically related individuals—parents and children, siblings, and so forth. The example I gave of worker ants helping the queen ant to survive and reproduce at the expense of their own reproductive prospects falls into this category. The worker ants, as is well known, are all sisters of the queen—all worker ants are female (Moffett 2010). Worker ants cannot have their own children, but it is thought that they ensure their genes are passed on by assisting in the survival and reproduction of their sister queen.

This alone cannot explain all cooperative behaviors in the animal kingdom, however. One major characteristic of human cooperation in particular is that it occurs so widely, even among genetically unrelated strangers. And so we come to the idea of **direct reciprocity** (Trivers 1971).

This is thought to be behind the example of vampire bat behavior that I gave. As the name suggests, vampire bats feed on the blood of animals, and they sometimes share this precious food with their roost-mates via regurgitation. And, moreover, this behavior is not limited to kin relationships. Researchers studying these bats reported that individuals have a propensity to share blood with those who had previously shared blood with them after they had failed to feed on a particular night. That is, they have a tendency to remember who helped them in the past and return the favor. This sort of direct helping is called direct reciprocity, and, as may be evident, it can explain cooperative behavior between two individuals that encounter each other repeatedly.

But, as you may be quick to point out, such direct, mutual exchanges of kindness are not the only type of cooperation in the human world. This brings us to the notion of **indirect reciprocity** (Nowak and Sigmund 1998, 2005).

Indirect reciprocity encapsulates the idea behind sayings like "one good deed deserves another" and "good karma comes around". It's the idea that if you help others out, then at some point someone else will help you out when you need it. Evolutionary biologist Martin A. Nowak, who studies cooperative behavior in humans, identifies indirect reciprocity as one of the important mechanisms to consider when thinking about cooperation among humans, and explains that reputation, which we have discussed in this book, is what underlies this mechanism (Nowak 2012). This is because those with a good reputation are likely to find themselves on the receiving end of goodwill when they are in need. Note that Nowak identifies five mechanisms for the evolution of cooperation (Nowak 2006). In addition to kin selection, direct reciprocity, and indirect reciprocity, which we have just discussed, his list also contains network reciprocity and group selection. For an in-depth discussion, see Nowak 2012.

# Afterword

My interest in the problems discussed in this book dates back to my own days as a student.

I was not an economics student at the time. The vicissitudes of life in general meant that I hadn't had the most diligent of high school careers, and I had entered university without really considering what direction I wanted to go in. I was far from a model student. I wasn't inspired by learning at all, and instead, by the time I realized it, had allowed myself to become wrapped up in student society activities (Singing in choirs!). At some point, I began to have doubts about my situation. Something had to change, and so throwing caution to the wind, I decided to attend courses at what was considered the best university in Japan. I just looked through the syllabus for courses that seemed even remotely interesting and attended as many as I could get into, irrespective of whether they were part of the sciences or the humanities. The one class I never skipped, the one that always held my attention all the way through, was microeconomics. I remember sitting in a large lecture hall, always craning forward, just a little from the left in the front row on the upper seating level. I was captivated by the beauty of economic theory and drawn in by the exhilarating insights of game theory. When I returned to my own university, that momentum carried me through. I went from spending days alone in the library to, eventually, taking on graduate school.

Feeling that I did not have a solid undergraduate foundation, I began my graduate school career with a sense of restlessness. I spent hours and hours ardently trying to comprehend the texts and papers I was reading, brushing aside the occasional simmering doubts I had about economics as stemming from my own lack of understanding.

I was fortunate enough to have the opportunity to study at the graduate school of economics at the University of Pennsylvania thereafter. There, my fascination with economics and game theory deepened even further. My time as a graduate student in Japan and the US was both demanding and fun, a jumble of elation and tribulation,


K. Kobayashi, *The Mystery of the Ultimatum Game*,
https://doi.org/10.1007/978-981-97-8979-5

an invaluable period in my life that I cannot describe in words, but one that I will never forget.

Shortly after returning to Japan, I came across the phrase, "a 'revolution' is taking place in economics" an academic conference of economics. I don't remember the exact words, but the claim that great changes were afoot in economics really grabbed me, marking the beginning of my time back in Japan.

Before long, I had taken on a teaching role myself, but those simmering doubts still rolled around in my head from time to time. At some point, it became clear to me that I would have to face them head on, and that's when I remembered those words from the conference. So I ventured into the pages of a behavioral economics textbook, and so quickly was I drawn in, nay, swallowed up by the topic that I surprised even myself. It was the first time I had been so captivated by learning something new since I encountered economics at the age of 20.

Even so, the more I learned about behavioral economics, the more I developed simmering doubts about it as well. A fellow faculty member of mine had very aptly described the subject as a "catalog of irrational behavior", and this really is what I was uneasy about—the tendency in some quarters to single out human irrationalities as some sort of amusing curiosity. I started to have doubts about this picture of humans as basically just simpleminded fools.

At the same time, I also felt uncomfortable about the tendency to focus exclusively on anomalies. When an experiment yields nothing but rational outcomes of the sort that economic theory would predict, people say, "Well, that's just too obvious. Where's the anomaly? There's no paper in this." Yet if all you do is catalog anomalies, of course you're going to end up with nothing more than an ever more hefty catalog of irrational behavior.

Brushing aside rational outcomes as "obvious" is, if you turn it around, tantamount to acknowledging that humans often behave like Econs. What I came to think is that if you only recognize rational behaviors as obvious, and if you fail to look, in totality, at both behaviors that seem rational and those that don't, then you have no chance of figuring out what separates the two, and, moreover, you will have a hard time figuring out what lies at the root of human choices. These little doubts simmering through my mind about traditional economics as well as behavioral economics, which arose as a critique of it, were the impetus for this book.

It was through my study of behavioral economics that I encountered evolutionary psychology. Evolutionary psychology considers issues of the mind through an evolutionary lens. To be a little grandiose about it, each step in my journey—when I encountered economics, and then behavioral economics, and then evolutionary psychology—changed the way I saw the world. Economics not only taught me *how* to approach and think about problems, it also instilled in me an understanding of the importance of the ability to think about problems at all, an ability on which everything else arguably rests. Behavioral economics taught me that it is okay to express my doubts and, indeed, to explore them. And evolutionary psychology gave me a completely new way of looking at the world.

In learning about evolution, I was deeply moved by how earnest and unrelenting the human brain, body, and mind are when it comes to surviving and carrying life forward. Life, we think, began some four billion years ago, back in a time beyond all memory, in the unimaginably distant past. All those countless years of struggling and striving to survive, to propagate life, are what led to us being here today. That the marks of that struggle are etched into us is an unmistakable fact; and if our own brains and bodies and minds are so driven to live, then we must live in a way that does all of this justice, and we must pass this knowledge on to the next generation—these are the thoughts that evolution inspired in me.

Learning and thinking bring nothing but pure joy to me these days. Being able to approach research and teaching with such a sense of excitement and wonder, even after all these years, is something I am so deeply grateful for. Two things are always at the forefront of my mind when I teach: the first is conveying the joy of learning and thinking, and the second is giving the students at least one "intellectual souvenir" to take home with them. There was another thought driving me as I wrote this book, one that I think reflects my own journey and just how much I had to learn on my own. I thought, I have to make this book easy to understand, it has to be clear and accessible, no matter what. The words that kept rolling around my head as I wrote came from someone I consider a senior of mine: "Think deeper than the ocean!".

This book tackles a wide range of topics, transcending the traditional boundaries between the sciences and the humanities. To enable discussions that cross disciplinary lines, I strongly felt that the book should be written in a way that anyone, regardless of their field of expertise, could easily understand. This is because, in order to truly understand human beings and to envision a society where everyone can live happily, it is essential to engage in conversations that go beyond any single discipline. To achieve this, I believed it was crucial to use language that is clear and accessible to everyone. As such, my goal was to create an "academic book for everyone,"—a book not only for researchers but also for a broader audience, including undergraduates and even high school students.

It has, as a consequence of this, however, become littered with footnotes, so much so that it's almost like two books in one: one half for a general readership and the other for the experts. I apologize to anyone who found it tough going for that reason. But, if you've made it this far, and if you've been able to take anything at all away from this book—if I've been able to give you at least one "intellectual souvenir" to take home—then I'm more than happy. And to my young readers in particular, I hope I've given you a taste of the unbridled joy that awaits in the single-minded exploration of those little simmering doubts and questions you have about the world around you. Go forth and discover!

Last but not least, I have a long list of people to thank for reading my early drafts of this book and providing invaluable feedback: Masahiro Okuno-Fujiwara at the University of Tokyo who supervised my graduate work, Toshiji Kawagoe at Future University Hakodate, Akihiko Matsui, Tatsuya Kameda, and Kazuya Kamiya at the University of Tokyo, Fumio Ohtake at Osaka University, Motonari Kurasawa at Yokohama National University, Tatsuhiko Nariu at Kyoto University, Nobuyuki

Kawai and Makoto Hanazono at Nagoya University, Ryo Oda at the Nagoya Institute of Technology, Yosuke Ohtsubo at Kobe University, Tadashi Sekiguchi at Kyoto University, Hisashi Ohtsuki at the Graduate University for Advanced Studies, Yoichi Hizen, Nobuhiro Mifune, and Toshiyuki Himichi at Koichi University of Technology, Hirofumi Hashimoto at Yasuda Women's University, Ryo Okui at Seoul National University, Satoshi Kawanishi at Sophia University, and Kaoru Ueda, Masanori Urakami, Yoshiaki Kaoru, Taro Okuda, Akira Nishimori, and Takeshi Goto at Nanzan University.

I received feedback and advice from so many experts from across such a wide range of fields, including economics, psychology, cognitive science, neuroscience, biology, ethics—the list goes on. Not only did this feedback help me in making vital improvements and corrections to the book, it also provided me with a whole lot of new insight and knowledge. Please allow me to take this opportunity to express my sincere gratitude to all of my reviewers.

I would like to offer my special thanks to Toshiji Kawagoe, Akihiko Matsui, and Yohsuke Ohtsubo for all of the back and forth via email. Not only were their comments apt, they also taught me a lot about writing, how to approach things as a researcher, and so much more. Fumio Ohtake not only provided excellent feedback and advice but was also very considerate on a personal level. Akira Nishimori, Masanori Urakami, Nobuhiro Mifune, Toshiyuki Himichi, and Yohsuke Ohtsubo read my manuscript twice, provided detailed, insightful feedback, brought relevant research papers to my attention, and, last but not least, offered heartwarming words of encouragement. I would like to thank Akira Nishimori, in particular, for responding thoughtfully to my incessant questions and for providing incisive yet warm comments at just the right time. And above all, I would like to thank Masahiro Okuno-Fujiwara, who offered insightful feedback on so many occasions and all sorts of encouragement along the way, and without whom I would never have been able to write this book.

I received guidance and insightful comments on topics related to my manuscript from Michihiro Kandori and Shozo Ota at the University of Tokyo, and Haruo Oyatsu, Yuuji Yumoto, Yoriko Hayashi, Yoshinori Okada, Takashi Ieda, and David Courron at Nanzan University. And, finally, I would like to acknowledge Masaru Kawasaki formerly of Nanzan University and Toshio Sakamoto at Nanzan University for creating the opportunity for me to publish this book in the first place.

Listing all of these names has reminded me of just how many people have assisted me along the way. I could never have written a book like this without them. They have my sincerest and heartfelt gratitude.

Makoto Muto, formerly of Nippon Hyoron Sha Publishers, and Fukiko Konishi at Nippon Hyoron Sha Publishers were a pleasure to work with. They provided insightful comments and advice and much warm encouragement. This book would not exist without them. Thank you both very much. I am also grateful to the illustrator Emiko Fuchigami for the lovely illustrations featured in these pages and on the book's cover. My heartfelt gratitude also goes out to the library staff who scoured libraries throughout Japan to collect resources on subjects such as the brain, infants, monkeys, Sahelanthropus tchadensis, and so on, which seemingly had nothing to do with economics. I would also like to again thank everyone from my seminar sessions who read the manuscript and provided valuable comments, as well as all the students who have ever taken my classes.

The process of writing this book has reminded me of the value of human connections, and with that in mind, I would also like to thank each single person I have had the pleasure to meet throughout my journey in economics.

I never realized how demanding it is to write a book. Having become a mother, I was already finding it difficult to attend any academic meetings and conferences, and gone were the days of freely setting aside great chunks of time for whatever I wanted to work on. After much agonizing, I figured that I just had to do whatever I could with the fleeting moments available to me here and there throughout the day. Much of this time was spent devouring research papers and books.

The more I read, the more literature I found I had to consume. I would read while preparing meals and end up burning the pot. I would read papers while briefly ensconced in a wardrobe or whatnot during a game of hide-and-seek. I would read while breastfeeding and drift off to sleep; this happened all the time. And before I knew it, I had well in excess of a thousand pages of notes.

Once I decided to put some part of them into a book, I was constantly drawn between all that research literature and the draft itself. I would read it and reread it, forever finding little things that I just had to tweak, change, revise, fix. Space limitations also meant that my biggest problem was not figuring out what to put in but, rather, figuring out what to cut. Every now and then, ideas popped into my barely conscious mind in the middle of the night, since it was the only quiet time of the day. Allowing myself to drift off would mean forgetting it entirely, and so I was compelled to rise and commit it to the page. Breastfeeding and motherhood alone were enough to constantly disturb my sleep, and the gnawing of inspirations for the book on top of that were driving me to my physical limits.

The breadth of subjects I was covering made it challenging at first to work out a scheme for consistent terminology usage throughout the book. I would spend days reviewing piles of documents just to settle on a convention for a few syllables. And yet I still worry to this day that the text is littered with technical blunders and the like that some expert somewhere will roll their eyes at. To all of my dear readers: if you notice anything out of place, please let me know. The blame for any errors and inaccuracies of course lies solely with me as the author.

Allow me to end on a personal note. While writing this book, I really struggled at home. My children were constantly admonishing me for working too hard, and I know they sorely missed the fully present and attentive me. My eldest son would look in on me whenever I was working and say a few words of encouragement. My second son was full of concern for me and constantly asking if the bookstore was going to be so good as to buy my book. And my daughter, the youngest of the family, added the word *rombun* (Japanese for research paper) to her budding vocabulary. The little darling would collect up any papers I'd left lying around the house and present them to me exclaiming, "Mama, rombun!", "Mama, your work". My father read through the entire manuscript and offered straightforward feedback as a layman, pointing out "I don't get this part!", "I don't get that part either!". My mother was always concerned about me and came to help me at home. My children ate their dinner and declared it delicious even when I'd burnt it. And my husband single-handedly did everything around the house, and even took care of that burnt pot without saying a word. To my dearest family, and to everyone else who has supported me through this, let me thank all of you from the bottom of my heart!

# Afterword to the English Edition

This book was originally published in Japan. When I first presented the ideas covered in this book at a seminar, my older children were 4 and 2 years old, and my youngest was not yet even born. Now, as this English version is set to be published after the Japanese edition, my oldest son is about to graduate from junior high, and my younger children are in elementary school. During the time, I also had to go through family emergencies, which repeatedly made me consider abandoning the publication. I was just so surprised at the length of time it took.

I feel deeply grateful to have been honored with the Nikkei Prize for Economics Books (the 64th Nikkei-Keizai Tosho Bunka award) for the original version in Japan, an award for outstanding economics books published in a given year. This gave me the opportunities to forge invaluable connections and gain new insights that extend well beyond the realm of economics.

At the award ceremony, a newspaper reporter asked me the following question: "Reading your book made me wonder about what human happiness actually is. What are your thoughts?".

This question prompted a memory of something my friend had said to me while I was an undergraduate student.

"Hey, Kayoko, I just think all humans were born to be happy."

When I heard these words, I thought, well that's blatantly obvious. Everyone wants to be happy and everyone strives to be happy. So saying that we were born to be happy is not really saying much of anything at all. Of course we are, is what I thought. That's like pointing out that the sky is blue.

I don't quite recall what I said to my friend at the time. Yet even though I thought it was obvious, it stuck with me for some reason, and I've never forgotten those words. It's the memory that sprang to mind when the reporter asked me that question.

And then it occurred to me that perhaps my friend and I were both wrong.

Evolution doesn't seem to care about our happiness. What it does care about is getting an individual to pass on as many of its genes as it can. Whether the individual lives a happy life or not does not factor into it.



K. Kobayashi, *The Mystery of the Ultimatum Game*,
https://doi.org/10.1007/978-981-97-8979-5

Evolutionary psychologist Robert Kurzban put it this way: "Evolution's interest in your happiness is—has to be—purely instrumental. It's not interested in making you happy for its own sake. Evolution works by leading to the development of systems that motivate you to do adaptive things—and avoid maladaptive ones like sticking pencils up your nose—with no regard for how happy or miserable that makes you on balance." (Kurzban 2011, p. 208).

But in that case, I thought, we're free of happiness. In other words, we are free to pursue our own path toward whatever it is we think of as happiness. If evolution has no interest in it, then the only person who really needs to take responsibility for your own happiness is you. If evolution has no position on what constitutes happiness, or misery for that matter, then you should simply go for whatever you think happiness is without hesitation (although, too much focus on happiness may not be good either...... (Mauss et al. 2011; Ford et al. 2014, 2015; Takai et al. 2023; Krys et al. 2024)).

That's how I answered the newspaper reporter's question. Yet later, as I continued to ponder the word "happiness," I came to realize that it wasn't just that.

When is it that we feel happy? Research has shown that we humans feel happy when we do something for someone else (Dunn and Norton 2013; Dunn et al. 2014), that we feel happier when we spend money on someone else rather than on ourselves (Dunn et al. 2008), that helping others can relieve physical pain (Wang et al. 2020), that wishing others well can improve mental state (Gentile et al. 2020), and that doing something nice for others can increase not only well-being but physical health as well (For a meta-analysis, see Hui et al. 2020). Showing kindness toward someone else need not be anything grandiose, either. Giving blood or volunteering are of course great examples, but even trivial acts, like showing a little kindness to a stranger or expressing gratitude to someone, make us feel happy too.

A recent study also found that people who were kind to others experienced less depression, anxiety, and loneliness compared to those who "rewarded" themselves through self-kindness. What's surprising is that, depression and anxiety were rather increased in those who were kind to themselves, even though they reported enjoying the act of "treating yourself" (Naclerio et al. 2024).

Forming healthy social relationships with others also seems to increase people's happiness levels (Dunbar 2021). And that's not all. Having healthy connections with others has repeatedly been shown to improve physical and mental health and even extend lifespan (Cacioppo and Patrick 2008; Wang et al. 2023).

A number of studies indicate that young children also show a remarkable tendency to be kind to others and to try to help them (Warneken and Tomasello 2006, 2007). Remarkably, children are willing to help even when doing so incurs a cost (Svetlova et al. 2010), even when it is not beneficial to the helper (Warneken and Tomasello 2013), or in cases where the recipient is unaware of being helped (Warneken 2013). Research further suggests that their joy comes not from any personal gain or sense of achievement but from simply seeing others freed from hardship (Hepach et al. 2012). All of these findings urge me to describe their behavior as genuine altruism; observations like these truly make me feel warm inside.

In this book, I've argued that the workings behind the seemingly altruistic act of sharing something of value with a stranger have something to do with superficial altruism driven by an innate concern for their own reputation, a somewhat disheartening revelation. As unfortunate as it may be, we cannot escape the fact that this is part of our nature.

But hang on a minute. Is that really all there is to us? I just told you about research suggesting that young children have an innate capacity for something I think is worthy of the name selfless altruism, and about research indicating that exhibiting altruistic behavior makes people happy. We also derive happiness from having healthy connections with others, and this in turn improves our physical and mental health.

The finding that we derive subjective happiness from doing something for others is, if you think about it, quite strange from the perspective of humans as living creatures. I think it's safe to say that, academically at least, there is still much we do not understand here. Even so, I think we can in fact say that we have an innate mechanism for deriving happiness from altruistic behavior.

Successfully slipping under everyone else's radar, evading those watchful eyes, and claiming whatever gains are up for grabs all to yourself does, undoubtedly, yield a peculiar kind of thrill and satisfaction. But is that happiness? Isn't happiness really being able to say to yourself with complete honesty that you've helped someone or been of use to them, even at a cost, or a loss, to yourself? And it can be something even more trivial. Imagine walking down a narrow pathway and momentarily stepping aside to give someone else room to get by. How do you feel when they shoot you a gleaming "Thank you!" as they pass? Don't fleeting interactions with strangers like this just warm your heart? Selfless altruism may sound quite extravagant, but it really can be just the little things. That's what makes us happy.

And I do believe that we humans have the power of reason. Creatures though we may be, are we humans not creatures of choice? We can choose whether to go for that thrill of self-serving satisfaction, or to instead seek the satisfaction of a heart-warming deed.

Of course, I think it is evident from the research on free will in recent years that the question of what "reason" actually is remains to be resolved. Still, I personally believe that we humans have something worthy of being called reason within us, that we have the power to choose. Will you choose the path of self-serving satisfaction? Or will you choose the path of warmth and happiness?

I'm no longer sure whether my friend who talked about happiness was right or not, but I think we are starting to gain some insight into what sort of paths we should take to live happily, and thinking only about yourself is certainly not the way to go.

Evolution apparently has no interest in your happiness, as I've said. Perhaps the only one responsible for deciding what happiness is to you, is you. I know as much as I hate to say that it is difficult, but still, if I could choose the "path to happiness" ……, that is what I am now thinking about. Which path will *you* choose……?

Now, let me wrap up with some background about this English edition. It is not a mere translation of the original version; rather, I have incorporated updated insights, removed certain sections too specific to Japan, excluded studies subject to doubts about reproducibility, incorporated some of the results of my research, reorganized the discussion, and provided enhanced clarity on my own academic arguments. Quantifying the magnitude of these changes is not easy, but in terms of the number of references cited, for instance, which was over 500, around 100–150 (or even more) have been replaced.

This book constructs its arguments based on the option chosen by the greatest number of participants in the studies cited, with limited exploration of the effects of factors such as diversity or culture. Nonetheless, it's crucial to clarify that this doesn't mean I consider those factors to be unimportant. While this book covers a multitude of experiments, hardly any of these experiments produced results indicating that almost all participants chose the same option. That is why, with the aim of conveying that fact to some extent, I made an effort to explicitly state the results of the major experiments cited, indicating how many of the subjects chose a particular option.

The original version faced harsh criticism, especially from the economics community, with some arguing that it isn't even economics to begin with, while others found that the sudden mention of evolution makes no sense at all. At first, I was striving to address each criticism, but over time, I've come to feel that those voices no longer matter. What I want to do is unravel the mysteries of human beings and create a society where every single *beings* can live happily together. It really doesn't matter whether we call the quest necessary for that purpose economics, psychology, comprehensive human science, or anything else. Just as is the case with happiness, I now believe that all I need to do is to courageously go for whatever I think is important, with sincerity and without any hesitation.

Last but not least, let me express my heartfelt thanks to everyone who helped me publish this book; Akira Nishimori, Taizo Motonishi, Nobuhiro Mifune, Masanori Urakami, Taro Okuda, Masahiro Okuno-Fujiwara, Akihiko Matsui, Shozo Ota, Takeshi Asamizuya, Michihiro Kandori, Akira Yakita, Kimiko Terai, Sadaharu Morinaka, Tami Yanagisawa, Toru Nagata, Fukiko Konishi, Emiko Fuchigami, Kiyoshi Fujimoto, Yoshiaki Kaoru, Nozomi Shimotori and all members of the Educational Planning and Research Promotion Division at Nanzan University, Eizaburou Kobayashi, Shizuko Kobayashi, Deepa Varadharajan, and their entire team. In particular, Juno Kawakami of Springer/Nature provided unwavering support from the inception of the idea to publish this book until its completion.

I also want to convey my genuine admiration and appreciation to Murdoch MacPhee and Mark Stevenson for their diligent work. Not only did they translate and/or proofread the English, but they also took the time to revisit each and every one of the extensive references to verify their content. Interestingly enough, Murdoch, introduced by Springer/Nature as the most reliable Japanese-English translator in this field, turned out to be a friend of mine from my days at the graduate school of economics. I am still at a loss for words to describe the surprise and joy I felt at our miraculous reunion.

Additionally, I extend my sincere gratitude to the Nitto Foundation, Takashima Kunio Jiyu Prize, Encouragement Award 2024, the Nanzan University Monograph Series, the Nanzan University Pache Research Subsidy 1-A-1 for the 2023 academic year, and the Nanzan University Pache Research Subsidy I-A-2 for the 2024 academic year, research grant from Murata Science and Education Foundation for their financial support for this publication.

As we approach the real final moments, allow me to share a personal note here once again. During the time I was working on the original version, I eagerly looked forward to spending more time with my children after its completion. Looking back, that period coincided with the global COVID-19 pandemic, which brought challenges for everyone. However, the response to the publication was much greater than I ever imagined, and before I knew it, I found myself not just busier than ever, but also needing more time to complete this English version. To my regret, this has made my children feel even lonelier at a time when they needed me most. To my dearest children, I want to express my heartfelt apologies and my deepest love. Please never forget: you are my greatest treasures.

Let me also express my heartfelt gratitude—8888 times—to my father and mother, whose unwavering love, encouragement, and support have been with me every step of my life. Your love and the strength with which you have approached life's journey are deeply etched in my soul. I am profoundly grateful to you, as well as to my brother and sisters, for your constant support, understanding, and heartfelt care.

And to my husband, who has always been there for me and our children, I want to express my deepest gratitude. I look forward to sharing time with you, not just now, but long beyond sixty-four!

*"Nothing in Biology Makes Sense Except in the Light of Evolution"*
*by Theodosius Dobzhansky*

To my loving family and children.

# Additional Details on Game Theory Games

This book has dealt primarily with the ultimatum and dictator games, but there are other game theory games out there. This section provides additional details on some particularly notable ones mentioned in this book.

1. **The public goods game**

We looked at the public goods game (with punishment opportunities) in our discussion of altruistic punishment in Sect. 4.2. The **public goods game** is used to investigate situations in which members of a group must each decide how much money they are willing to contribute to the creation of a public good, such as a park or a road. The players are free to decide how much they will individually contribute to the public good.

We'll consider the following experiment setup. The players are put into groups of four, and at the start of every game, each of the four players is given 20 money units (MUs). Each player then decides how much of that private stake they will contribute to a public pot, which will be used for everyone's benefit. The public pot is used to build public goods, like parks and roads, and the money contributed by each player can be thought of as taxes.

The researcher multiplies the money contributed to the public pot by some factor—the researcher could double the money held in the public pot, for instance. The resulting amount is then distributed evenly among the players. Hence, the most beneficial situation for the group as a whole is for every player to contribute their entire private stake to the public pot. The researcher would then (in our example) double the money and return it to the players, and this results in the maximum possible total payout for the group as a whole.

For each individual, however, the most beneficial option is to contribute no money at all. Whatever you contribute will be (again, in our example) doubled by the researcher and returned to you, but because the pot is divided evenly among the group members, you only stand to receive a quarter of what the pot holds. If, as in our example, the researcher doubles the money in the pot, this means that you


K. Kobayashi, *The Mystery of the Ultimatum Game*,
https://doi.org/10.1007/978-981-97-8979-5

will receive back a mere half of what you just put in (plus your share of the other contributors' sums). So the best situation for you as an individual is for all three of the other players to contribute their entire private stake while you chip in nothing at all.

Everyone may freely use parks and roads created using tax money, regardless of how much tax they individually paid. If you can have parks and roads built using money contributed by others without having to pay a single cent yourself, that would provide the greatest benefit for you as an individual. But if everyone thought that way and no one paid taxes, we would not be able to build any roads or parks in the first place. The public goods game is intended to represent this aspect of reality.

Incidentally, the experiment cited in this book used a multiplication factor of 1.6, which means that any individual stood to receive back 0.4× their initial contribution (plus their share of the others' contributions). In the paper, above-average contributors are referred to as cooperators, and below-average contributors are referred to as defectors. The degree of cooperation is measured by how much of the initial 20-MU stake each individual contributed.

A punishment stage was also added to the game, giving the players the opportunity to engage in altruistic punishment (costly second-party punishment). The players were informed of all other group members' contributions, and then allowed to punish each of the other group members by having them pay a cost. In this punishment stage, each player was able to assign 0–10 points to the punished member. Each point assigned cost the punished member 3 MUs and the punishing member 1 MU.

2. **The sequential prisoner's dilemma game**

The game used to measure people's joy at seeing others in pain, as discussed in Sect. 4.3, is called the **sequential prisoner's dilemma game**.

This is a game played by two people: the first mover and the second mover. Each player receives 10 points to start with. The first mover decides how much of his 10 points to give to the other player (the second mover). Each point sent to the second mover is tripled upon receipt. After seeing how many points the first mover has given to her, the second mover then decides how much of her initial 10 points to give to the first mover. Each of these points is also tripled upon receipt. This ends the game.

In the experiment cited in this book, the experimental subjects always played the part of the first mover. The second movers were all confederates (individuals posing as participants who were actually collaborating with the researcher). As explained in Sect. 4.3, there were two types of confederates: fair ones, who returned many points, and unfair ones, who returned few points. The researchers then measured the brain response in experimental subjects when seeing the unfair "player", who was trying to hog the money, experience a pain stimulus.

3. **The trust game**

The game used in the discussion of the joy of punishing others in Sect. 4.3 is called the **trust game**. This is a game played by two people. Each player receives 10 points to start with.

First, Player 1 decides how many of his endowment of 10 points to send to Player 2. Each point sent is quadrupled by the researcher upon receipt. Hence, if Player 1 sends 10 points, for example, then Player 2 receives 40 points. At this point, Player 2 would have a total of 50 points (initial endowment plus points received from Player 2).

Player 2 then decides how many of these 50 points to send back Player 1. Once Player 1 receives those points, the game ends. The points each player has in hand at the end are their respective gains.

The best thing for the pair collectively is, of course, for Player 1 to send all of his 10 points to Player 2. Those 10 points turn into 40, giving the greatest possible overall gain. The best thing for Player 2 individually, however, is to then refuse to send anything back to Player 1. Player 2 receives no additional gain by sending anything back to Player 1. So, if you can predict this outcome from the beginning, it is evident that the best thing for Player 1 to do as an individual is to refuse to send any points to Player 2 to start with.

After sending some of his points to Player 2, Player 1 has no idea how many points, if any, will be returned, so his behavior can be seen as a measure of how much he trusts Player 2, and hence the name of the game.

In the experiment cited in this book, the experimental subjects always played the part of Player 1. The part of Player 2 was played by confederates (individuals posing as participants who were actually collaborating with the researcher). There were two types of confederates: trustworthy ones, who validated Player 1's trust by returning points, and defectors, who abused Player 1's trust and kept the points to themself. The researchers then measured the brain response in experimental subjects when a punishment was administered to the unfair "player" who had abused the person's trust and pocketed the points for themself.

*The author utilized ChatGPT and DeepL as tools for translation when rewriting from the Japanese version to the English version.*

# References


Alicart, H., D. Cucurell, & J. Marco-Pallarés (2020) "Gossip information increases reward-related oscillatory activity,"***NeuroImage***, 116520.

Ames, D. L., & Fiske, S. T. (2013), "Intentional harms are worse, even when they're not," Psychological science, 24(9), pp. 1755–1762.

Andersen, S., S. Ertac, U. Gneezy, M. Hoffman, & J. A. List (2011) "Stakes matter in ultimatum games,"***American Economic Review***, 101(7), pp. 3427–3439.

Anderson, C. M., & L. Putterman (2006) "Do non-strategic sanctions obey the law of demand? The demand for punishment in the voluntary contribution mechanism,"***Games and Economic Behavior***, 54(1), pp. 1–24.

Anderson, E., E. H. Siegel, E. Bliss-Moreau, & L. F. Barrett (2011) "The visual impact of gossip,"***Science***, 332(6036), pp. 1446–1448.

Anderson, J. R., B. Bucher, H. Chijiwa, H. Kuroshima, A. Takimoto, & K. Fujita (2017) "Third-party social evaluations of humans by monkeys and dogs,"***Neuroscience & Biobehavioral Reviews***, 82, pp. 95–109.

Anderson, J. R., H. Kuroshima, A. Takimoto, & K. Fujita (2013a) "Third-party social evaluation of by monkeys,"***Nature Communications***, 4, 1561.

Anderson, J. R., A. Takimoto, H. Kuroshima, & K. Fujita (2013b) "Capuchin monkeys judge third-party reciprocity,"***Cognition***, 127(1), pp. 140–146.

Andrea, L., R. Amanda, & B. Pat (2019) "Stake size effects in ultimatum game and dictator game offers: A meta-analysis,"***Organizational Behavior and Human Decision Processes***, 151 (C), pp. 61–72.

Andreoni, J. (2007) "Giving gifts to groups: How altruism depends on the number of recipients,"***Journal of Public Economics***, 91(9), pp. 1731–1749.

Andreoni, J., & B. D. Bernheim (2009) "Social image and the 50–50 norm: A theoretical and experimental analysis of audience effects,"***Econometrica***, 77(5), pp. 1607–1636.

Andreoni, J., & J. Miller (2002) "Giving according to GARP: An experimental test of the consistency of preferences for altruism,"***Econometrica***, 70(2), pp. 737–753.

Archer, J. (2004) "Sex differences in aggression in real-world settings: A meta-analytic review,"***Review of general Psychology***, 8(4), pp. 291–322.

Artavia, L-M., A. S. Bedi, & M. Rieger (2017) "Intuitive help and punishment in the field,"***European Economic Review***, 92, pp. 133–145.

Ayal, S., J. Celse, & G. Hochman, (2021) "Crafting messages to fight dishonesty: A field investigation of the effects of social norms and watching eye cues on fare evasion,"***Organizational Behavior and Human Decision Processes***, 166, pp. 9–19.

Baillon, A., A. Selim, & D. van Dolder, (2013) "On the social nature of eyes: the effect of social cues in interaction and individual choice task,"***Evolution and Human Behavior***, 34(2), pp. 146–154.




K. Kobayashi, *The Mystery of the Ultimatum Game*, https://doi.org/10.1007/978-981-97-8979-5

Balafoutas, L., & N. Nikiforakis (2012) "Norm enforcement in the city a natural field experiment,"***European Economic Review***, 56, pp. 1773–1735.

Balafoutas, L., K. Grechenig, & N. Nikiforakis (2014a) "Third party punishment and counter punishment in one shot interactions," ***Economics Letters***, pp. 308–310.

Balafoutas, L., N. Nikiforakis, & B. Rockenbach (2014b) "Direct and indirect punishment among strangers in the field,"***Proceedings of the National Academy of Sciences***, 111(7), pp. 15924–15927.

Balafoutas, L., N. Nikiforakis, & B. Rockenbach (2016) "Altruistic punishment does not increase with the severity of norm violations in the field,"***Nature Communications***, 7, 13327.

Baldassarri, D. (2020) "Market integration accounts for local variation in generalized altruism in a nationwide lost-letter experiment,"***Proceedings of the National Academy of Sciences***, 117(6), pp. 2858–2863.

Ballard, T., D. K. Sewell, D. Cosgrove, & A, Neal, (2019) "Information Processing Under Reward Versus Under Punishment,"***Psychological Science***, 30(5), pp. 757–764.

Balliet, D., L. B. Mulder, & P. A. Van Lange (2011) "Reward, punishment, and cooperation: A meta-analysis, "***Psychological Bulletin***, 137(4), pp. 594–615.

Barclay, P. (2006) "Reputational benefits for altruistic punishment," ***Evolution and Human Behavior***, 27(5), pp. 325–344.

Barclay, P., & M. L. Lalumière (2006) "Do people differentially remember cheaters?,"***Human Nature***, 17(1), pp. 98–113.

Bardsley, N. (2008) "Dictator game giving: altruism or artefact?,"***Experimental Economics***, 11(2), pp. 122–133.

Barmettler, F., E. Fehr, & C. Zehnder (2012) "Big experimenter is watching you! Anonymity and prosocial behavior in the laboratory,"***Games and Economic Behavior***, 75(1), pp. 17–34.

Bartal, I. B. A., J. Decety, & P. Mason (2011) "Empathy and pro-social behavior in rats,"***Science***, 334(6061), pp. 1427–1430.

Batson, C. D. (2011) Altruism in humans, Oxford University Press.

Batson, C.D., E. R. Tompson, G. Seuferling, H. Whitney, & J. A. Strongman (1999) "Moral hypocrisy: Appearing moral to oneself without being so,"***Journal of Personality and Social Psychology***, 77, pp. 523–537.

Bateson, M., D. Nettle, & G. Roberts (2006) "Cues of being watched enhance cooperation in a real-world setting,"***Biology Letters***, 2, pp. 412–414.

Bateson, M., R. Robinson, T. Abayomi-Cole, J. Greenlees, A. O'Connor, & D. Nettle (2015) "Watching eyes on potential litter can reduce littering: evidence from two field experiments,"***PeerJ***, 3, e1443.

Baumard, N. (2010) "Has punishment played a role in the evolution of cooperation? A critical review,"***Mind and Society***, 9, pp. 171–192.

Baumard, N., O. Mascaro, & C. Chevallier (2012) "Preschoolers are able to take merit into account when distributing goods,"***Developmental Psychology***, 48(2), pp. 492–498.

Beaman, A. L., B. Klentz, E. Diener, & S. Svanum (1979) "Self-awareness and transgression in children: two field studies,"***Journal of Personality and Social Psychology***, 37, pp. 1835–1846.

Beersma, B., & G, A. Van Kleef (2011), "How the grapevine keeps you in line: Gossip increases contributions to the group," ***Social Psychological and Personality Science***, 2(6), pp. 642–649.

Bell, R., & A. Buchner (2009) "Enhanced source memory for names of cheaters,"***Evolutionary Psychology***, 7(2), pp. 317–330.

Bell, R., C. Schain, & G. Echterhoff (2014) "How selfish is memory for cheaters? Evidence for moral and egoistic biases,"***Cognition***, 132(3), pp. 437–442.

Bellucci, G., J. A. Camilleri, V. Iyengar, S. B. Eickhoff, & F. Krueger (2020) "The emerging neuroscience of social punishment: Meta-analytic evidence," ***Neuroscience & Biobehavioral Reviews***, 113, pp. 426–439.

Belot, M., R. Duch, & L. Miller (2015) "A comprehensive comparison of students and non-students in classic experimental games,"***Journal of Economic Behavior & Organization***, 113, 26–33.

Benedict, R. (1946) The chrysanthemum and the sword: Patterns of Japanese culture, Houghton Mifflin Harcourt.

Bereczkei, T., B. Birkas, & Z. Kerekes (2010) "Altruism towards strangers in need: costly signaling in an industrial society,"***Evolution and Human Behavior***, 31(2), pp. 95–103.

Bernhard, H., E. Fehr, & U. Fischbacher (2006) "Group affiliation and altruistic norm enforcement,"***American Economic Review***, 96(2), pp. 217–221.

Blaine, T., & P. Boyer (2018) "Origins of sinister rumors: A preference for threat-related material in the supply and demand of information,"***Evolution and Human Behavior***, 39(1), pp. 67–75.

Blake, P. R., & K. McAuliffe (2011) "I had so much it didn't seem fair; Eight-year-olds reject two forms of inequity,"***Cognition***, 120(2), pp. 168–175.

Bland, A. R., J. P. Roiser, M. A. Mehta, B. J. Sahakian, T. W. Robbins,, & R. Elliott, (2022) "The impact of COVID-19 social isolation on aspects of emotional and social cognition,"***Cognition & Emotion***, 36(1), pp. 49–58.

Bloom, P. (2013) "Just babies: The origins of good and evil," Crown Publishers/Random House.

Bloom, P. (2016) Against empathy: The case for rational compassion, Ecco.

Blount, S. (1995) "When social outcomes aren't fair: The effect of causal attributions on preferences,"***Organizational Behavior and Human Decision Processes***, 63(2), pp. 131–144.

Blum, D. (2002) Love at Goon Park: Harry Harlow and the Science of Affection, Basic Books.

Boehm, C. (2012) Moral origins: The evolution of virtue, altruism and shame, Basic Books.

Boesch, C., C. Bole, N. Eckhardt, & H. Boesch (2010) "Altruism in forest chimpanzees: the case of adoption,"***PLoS ONE***, 5(1): e8901.

Bolton, G. E., E. Katok, & R. Zwick (1998) "Dictator game giving: Rules of fairness versus acts of kindness,"International Journal of Game Theory, 27(2), pp. 269–299.

Bone, J. E., & N. J. Raihani, (2015) "Human punishment is motivated by both a desire for revenge and adesire for equality,"***Evolution and Human Behavior***, 36, pp. 323–330.

Bourrat, P., N. Baumard, & R. McKay (2011) "Surveillance cues enhance moral condemnation,"***Evolutionary Psychology***, 9(2), pp. 193–199.

Bouwmeester, S., P. P. Verkoeijen, B. Aczel, F. Barbosa, L. Bègue, P. Brañas-Garza,..., & C. E. Wollbrant (2017) "Registered replication report: Rand, Greene, and Nowak (2012),"***Perspectives on Psychological Science***, 12(3), pp. 527–542.

Bowlby, J. (1973) "Affective bonds: Their nature and origin,"in R. S. Weiss ed., Loneliness: The experience of emotional and social isolation, MIT Press, pp. 38–52.

Bowles, S. (2003) Microeconomics: Behavior, institutions, and evolution, Princeton University Press.

Bowles, S. (2012) The new economics of inequality and redistribution (Federico Caffè Lectures), Cambridge University Press.

Bowles, S., & H. Gintis (2013) A cooperative species: Human reciprocity and its evolution, Princeton University Press.

Bradley, A., C. Lawrence, & E. Ferguson (2018) "Does observability affect prosociality?,"Proceedings of the Royal Society B: Biological Sciences, 285 (1875), 20180116.

Bräuer, J., J. Call & M. Tomasello (2006) "Are apes really inequity averse?," ***Proceedings of the Royal Society B: Biological Sciences***, 273 (1605), pp. 3123–3128.

Broberg, T., T. Ellingsen, & M. Johannesson (2007) "Is generosity involuntary?,"***Economics Letters***, 94(1), pp. 32–37.

Brondino, N., L. Fusar-Poli, & P. Politi (2017) "Something to talk about: Gossip increases oxytocin levels in a near real-life situation,"***Psychoneuroendocrinology***, 77, pp. 218–224.

Brosnan, S. F. (2013) "Justice-and fairness-related behaviors in nonhuman primates,"***Proceedings of the National Academy of Sciences***, 110(Supplement 2), pp. 10416–10423.

Brosnan, S. F., & F. B. M. de Waal (2003) "Monkeys reject unequal pay,"***Nature***, 425, pp. 297–299.

Brosnan, S. F., C. Talbot, M. Ahlgren, S. P. Lambeth, & S. L. Schapiro (2010)"Mechanisms underlying responses to inequitable outcomes in chimpanzees, Pan troglodytes,"***Animal Behaviour***, 79(6), pp. 1229–1237.

Brown, D. E. (1991) Human universals, McGraw-Hill.

Brown, W. M., & C. Moore (2000) "Is prospective altruist-detection an evolved solution to the adaptive problem of subtle cheating in cooperative ventures? Supportive evidence using the Wason selection task,"***Evolution and Human Behavior***, 21(1), pp. 25–37.

Bruhin, A., K. Janizzi, & C. Thöni (2020) "Uncovering the heterogeneity behind cross-cultural variation in antisocial punishment,"***Journal of Economic Behavior & Organization***, 180, pp. 291–308.

Brucks, D., J. L. Essler, S. Marshall-Pescini, & F. Range (2016) "Inequity aversion negatively affects tolerance and contact-seeking behaviours towards partner and experimenter," ***PLoS One***, 11(4), e0153799.

Brucks D, & A. M. P. von Bayern, (2020), "Parrots voluntarily help each other to obtain food rewards," ***Current Biology***, pp. 292–297.

Bu, F., J. Abell, P. Zaninotto, & D. Fancourt (2020) "A longitudinal analysis of loneliness, social isolation and falls amongst older people in England,"***Scientific Reports***, 10(1), 20064.

Buchner, A., R. Bell, B. Mehl, & J. Much (2009) "No enhanced recognition memory, but better source memory for faces of cheaters,"***Evolution and Human Behavior***, 30(3), pp. 212–224.

Bueno-Guerra, N., C. J. Völter, A. de Las Heras, M. Colell, & J. Call, (2019) "Bargaining in chimpanzees (Pan troglodytes): The effect of cost, amount of gift, reciprocity, and communication," ***Journal of Comparative Psychology***, 133(4), 542.

Burkett, J. P., E. Andari, Z. V. Johnson, D. C. Curry, F. B. de Waal, & L. J. Young (2016) "Oxytocin-dependent consolation behavior in rodents,"***Science***, 351(6271), pp. 375–378.

Burnham, T. C. (2007), "High-testosterone men reject low ultimatum game offers," ***Proceedings of the Royal Society B: Biological Sciences***, 274(1623), pp. 2327–2330.

Burnham, T., & B. Hare (2006) "Engineering human cooperation: Does involuntary neural activation increase public goods contributions in adult humans,"***Human Nature***, 17, pp. 90–108.

Buss, D. (2019) Evolutionary psychology: The new science of the mind, sixth edition, Routledge

Cacioppo, J. T., & W. Patrick (2008) Loneliness: Human nature and the need for social connection, W. W. Norton & Company.

Cai, W., X. Huang, S. Wu, & Y. Kou (2015) "Dishonest behavior is not affected by an image of watching eyes, " ***Evolution and Human Behavior***, 36(2), pp. 110–116.

Call, J., & M. Tomasello (1998) "Distinguishing intentional from accidental actions in orangutans (Pongo pygmaeus), chimpanzees (Pan troglodytes) and human children (Homo sapiens),"***Journal of Comparative Psychology***, 112(2), pp. 192–206.

Call, J., & M. Tomasello (2008) "Does the chimpanzee have a theory of mind? 30 years later,"***Trends in Cognitive Sciences***, 12(5), pp. 187–192.

Callaisa, J., C. Harrisb, B. & Borchard, (2022), "The moral costs of markets: Testing the deterioration hypothesis," Journal of Economic Behavior & Organization, 204, pp. 200–220.

Camerer, C. F. (2003) Behavioral game theory: Experiments in strategic interaction, Princeton University Press.

Cappelletti, D., W. Güth, & M. Ploner (2011) "Being of two minds: Ultimatum offers under cognitive constraints,"***Journal of Economic Psychology***, 32(6), pp. 940–950.

Carbon, C. C., & V. M. Hesslinger (2011) "Bateson et al.'s (2006) cues-of-being-watched paradigm revisited,"***Swiss Journal of Psychology***, 70(4), pp. 203–210.

Carlsmith, K. M., J. M. Darley, & P. H. Robinson, (2002) "Why do we punish? Deterrence and just deserts as motives for punishment,"***Journal of Personality and Social Psychology***, 83(2), pp. 284–299.

Carlsmith, K. M., T. D. Wilson, & D. T. Gilbert (2008) "The paradoxical consequences of revenge,"***Journal of Personality and Social Psychology***, 95(6), pp. 1316–1324.

Carrillo, M., Y. Han, F. Migliorati, M. Liu, V. Gazzola, & C. Keysers (2019) "Emotional mirror neurons in the rat's anterior cingulate cortex,"***Current Biology***, 29(8), pp. 1301–1312.

Carter, G. G., & G. S. Wilkinson (2013) "Food sharing in vampire bats: Reciprocal help predicts donations more than relatedness or harassment,"***Proceedings of the Royal Society B: Biological Sciences***, 280 (1753), 20122573.

Cervone, C., C. Suitner, L. Carraro, A. Menini, & A. Maass, (2024), "Unequal by malice, protesters by outrage: Agent perceptions drive moralization of, and collective action against, inequality," ***British Journal of Social Psychology***.

Chapman, H. A., D. A. Kim, J. M. Susskind, & A. K. Anderson (2009) "In bad taste: Evidence for the oral origins of moral disgust,"***Science***, 323 (5918), pp. 1222–1226.

Charness, G. & M. Pingle (eds.) (forthcoming) The art of experimental economics: Twenty top papers reviewed, Routledge.

Cheng, X., L. Zheng, L. Li, Y. Zheng, X. Guo, & G. Yang (2017) "Anterior insula signals inequalities in a modified Ultimatum Game,"***Neuroscience***, 348, pp. 126–134.

Chernyak, N., K. L. Leimgruber, Y. C. Dunham, J. Hu, & P. R. Blake (2019), "Paying back people who harmed us but not people who helped us: Direct negative reciprocity precedes direct positive reciprocity in early development," ***Psychological Science***, 30(9), pp. 1273–1286.

Chiappe, D., A. Brown, B. Dow, J. Koontz, M. Rodriguez, & K. McCulloch (2004) "Cheaters are looked at longer and remembered better than cooperators in social exchange situations,"***Evolutionary Psychology***, 2, pp. 108–120.

Chijiiwa, H., H. Kuroshima, Y. Hori, J. R. Anderson, & K. Fujita (2015) "Dogs avoid people who behave negatively to their owner: Third-party affective evaluation,"***Animal Behavior***, 106, pp. 123–127.

Church, R. M. (1959) "Emotional reactions of rats to the pain of others,"***Journal of Comparative and Physiological Psychology***, 52(2), pp. 132–134.

Cikara, M., & S. T. Fiske (2012) "Stereotypes and schadenfreude: Affective and physiological markers of pleasure at outgroup misfortunes,"Social Psychological and Personality Science, 3(1), pp. 63–71.

Cochard, F., J. Le Gallo, N. Georgantzis, & J.-C. Tisserand (2021) "Social preferences across different populations: Meta-analyses on the ultimatum game and dictator game,"***Journal of Behavioral and Experimental Economics***, 90, 101613.

Colombatto, C., B. J. & Scholl (2022) "Unconscious pupillometry: An effect of "attentional contagion" in the absence of visual awareness," ***Journal of Experimental Psychology***: General, 151(2), pp. 302–308.

Cooper, D. J., & J. H. Kagel (2016) "Other regarding preferences a selective survey of experimental results,"in J. H. Kagel, & A. E. Roth (Eds.), The handbook of experimental economics Vol.2, Princeton University Press.

Cosmides, L. (1989) "The logic of social exchange: Has natural Selection Shaped with Humans Reason? Studies with the Wason Selection Task,"***Cognition***, 31(3), pp. 187–276.

Cosmides, L., & J. Tooby (1989) "Evolutionary psychology and the generation of culture, Part II: A computational theory of social exchange,"***Ethology and Sociobiology***, 10(1–3), pp. 51–97.

Cosmides, L., & J. Tooby (1997) "The multimodular nature of human intelligence,"in A. Schiebel, & J. W. Schopf (Eds.), Origin and evolution of intelligence, Jones & Bartlett Learning, pp. 71–101.

Cosmides, L., & J. Tooby (2005) "Social exchange: The evolutionary design of a neurocognitive system,"in M. S. Gazzaniga (Ed.), The Cognitive Neurosciences III, pp. 1295–1308, MIT Press.

Cosmides, L., H. C. Barrett, & J. Tooby (2010) "Adaptive specializations, social exchange, and the evolution of human intelligence,"***Proceedings of the National Academy of Sciences***, 107 (Supplement 2), pp. 9007–9014.

Cutler, J., J. P. Nitschke, C. Lamm, & P. L. Lockwood (2021) "Older adults across the globe exhibit increased prosocial behavior but also greater in-group preferences," ***Nature Aging***, 1(10), pp. 880–888.

Cummins, D. D. (1999) "Cheater detection is modified by social rank: The impact of dominance on the evolution of cognitive functions,"***Evolution and Human Behavior***, 20(4), pp. 229–248.

Cushman, F., K. Gray, A. Gaffey, & W. B. Mendes (2012) "Simulating murder: The aversion to harmful action,"***Emotion***, 12(1), pp. 2–7.

Dhaliwal, N. A., I. Patil, & F. Cushman, (2021) "Reputational and cooperative benefits of third-party compensation." ***Organizational Behavior and Human Decision Processes***, 164, pp. 27–51.

Dana, J., D. M. Cain, & R. M. Dawes (2006) "What you don't know won't hurt me: Costly (but quiet) exit in dictator games,"***Organizational Behavior and Human Decision Processes***, 100(2), pp. 193–201.

Dana, J., R. A. Weber, & J. X. Kuang (2007) "Exploiting moral wiggle room: experiments demonstrating an illusory preference for fairness,"***Economic Theory***, 33(1), pp. 67–80.

Darwin, C. (1871) The descent of man and selection in relation to sex, John Murray.

Dawkins, M, B., S. Sloane, & R. Baillargeon (2019) "Do infants in the first year of life expect equal resource allocations?,"Frontiers in Psychology, 10(116).

Dear, K., K. Dutton, & E. Fox (2019) "Do 'watching eyes' influence antisocial behavior? A systematic review & meta-analysis,"***Evolution and Human Behavior***, 40(3), pp. 269–280.

Decety, J., & K. J. Michalska (2010) "Neurodevelopmental changes in the circuits underlying empathy and sympathy from childhood to adulthood,"***Developmental Science***, 13(6), pp. 886–899.

de Quervain, D. J. -F., U. Fischbacher, V. Treyer, M. Schellhammer, U. Schnyder, A. Buck, & E. Fehr (2004) "The neural basis of altruistic punishment,"***Science***, 305(5688), pp. 1254–1258.

Dennett, D. C. (2006) Breaking the spell: Religion as a natural phenomenon, Viking Adult.

DeSteno, D. (2014) The truth about trust: How it determines success in life, love, learning, and more, Avery.

Deutchman, P., M. Bračič, N. Raihani, & K. McAuliffe (2021) "Punishment is strongly motivated by revenge and weakly motivated by inequity aversion,"***Evolution and Human Behavior***, 42(1), pp. 12–20.

de Waal, F. (2005a) "How animals do business,"Scientific American, 292(4), pp. 72–79.

de Waal, F. (2005b) Our inner ape, Reverhead Books.

de Waal, F. (2009) The age of empathy: Nature's lessons for a kinder society, Crown.

de Waal, F. (2013) The bonobo and the atheist: In search of humanism among the primates, W. W. Norton & Co. Inc.

de Waal, F. (2014) "One for all,"Scientific American, 311(3), pp. 68–71.

de Waal, F. (2019) "Empathetic behavior,"in Encyclopedia of Animal Behavior, 2nd ed., pp. 366–370, Academic Press.

Dobzhansky, T. (1973). Nothing in Biology Makes Sense Except in the Light of Evolution. American Biology Teacher, 35(3):125–129.

Dondi, M., F. Simion, & G. Caltran (1999) "Can newborns discriminate between their own cry and the cry of another newborn infant?,"***Developmental Psychology***, 35(2), pp. 418–426.

Dorison, C. A., & B. H. Heller, (2022), "Observers penalize decision makers whose risk preferences are unaffected by loss–gain framing,"***Journal of Experimental Psychology: General***. 151(9), pp. 2043–2059.

Dorison, C. A., C. K. Umphres, & J. S. Lerner, (2022) "Staying the course: Decision makers who escalate commitment are trusted and trustworthy,"***Journal of Experimental Psychology, General***, 151(4), pp. 960–965.

Drouvelis, M., (2023) "Dictator giving and taking: Evidence from Africa,"***Journal of Economic Behavior & Organization***, 209, pp. 326–333.

Dreber, A., D. G. Rand, D. Fudenberg, &M. A. Nowak (2008) "Winners don't punish,"***Nature***, 452 (7185), pp. 348–351.

Dunbar, R. (1996) Grooming, gossip, and the evolution of language, Farber & Farber.

Dunbar, R. (2004) "Gossip in evolutionary perspective,"***Review of General Psychology***, 8(2), pp. 100–110.

Dunbar, R. (2010) How many friends does one person need?: Dunbar's number and other evolutionary quirks, Harvard University Press.

Dunbar, R. (2011) "Evolutionary psychology in the round,"in Wang, X. T., (ed.) Thus spake evolutionary psychologists, pp. 23–31, Peking University Press.

Dunbar, R. (2014) "Human evolution,"Pelican Books.

Dunbar, R. (2021) "Friends: Understanding the power of our most important relationships."Little Brown

Eadeh, F. R., S. A. Peak, & A. J. Lambert (2017) "The bittersweet taste of revenge: On the negative and positive consequences of retaliation,"***Journal of Experimental Social Psychology***, 68, pp. 27–39.

Dunn, E., & M. Norton (2013) "Happy money: The science of happier spending," Simon and Schuster.

Dunn, E. W., L. B. Aknin, & M. I. Norton (2008) "Spending money on others promotes happiness," ***Science***, 319(5870), pp. 1687–1688.

Dunn, E. W., L. B. Aknin, & M. I. Norton (2014) "Prosocial spending and happiness: Using money to benefit others pays off," ***Current Directions in Psychological Science***, 23(1), pp. 41–47.

Eder, A. B., A. Krishna, & V. Mitschke, (2021) "Sweet revenge feels less bitter: Spontaneous affective reactions after revenge taking." ***Social Psychological and Personality Science***, 12(7), pp. 1184–1192.

Edgeworth, F. Y. (1881) An essay on the application of mathematics to the moral sciences, C. Kegan, Paul & Co.

Eidelman, S., C. S. Crandall, J. A. Goodman, & J. C. Blanchar, (2012) "Low-effort thought promotes political conservatism," ***Personality and Social Psychology Bulletin***, 38(6), pp. 808–820.

Eisenberger, N. I., M. D. Lieberman, & K. D. Williams (2003) "Does rejection hurt? An fMRI Study of Social Exclusion,"***Science***, 302(5643), pp. 290–292.

Eisenberger, N. I., S. E. Taylor, S. L. Gable,C. J. Hilmert, & M. D. Lieberman (2007) "Neural pathways link social support to attenuated neuroendocrine stress responses,"***Neuroimage***, 35(4), pp. 1601–1612.

Ekström, M. (2012) "Do watching eyes affect charitable giving? "Experimental Economics, 12, pp. 530–546.

Endo T. (2005) "Chapter 1: Review: The Origin of the Mind through Gaze Understanding: Seeing Eyes, Perceiving Gazes and Reading Minds', Reading Eyes, Being Read (sousetsu: sisen rikai wo tooshite miru kokoro no genryu: ganmoku wo miru · shisen wo utsusu · shini wo yomu)," The Evolutionary and Developmental Psychology of Gaze Understanding. (in Japanese) University of Tokyo Press, Endo T. (ed.)

Emler, N. (1994) "Gossip, reputation, and social adaptation,"in R. F. Goodman, & A. Ben-Ze'ev (Eds.), Good gossip, University Press of Kansas.

Engel, C. (2011) "Dictator games: A meta study,"***Experimental Economics***, 14(4), pp. 583–610.

Engelmann, J. M., & M. Tomasello (2019) "Children's sense of fairness as equal respect,"***Trends in Cognitive Sciences***, 23(6), pp. 454–463.

Enke, B., (2022), "Market exposure and human morality," ***Nature Human Behavior***

Ernest-Jones, M., D. Nettle, & M. Bateson (2011) "Effects of eye images on everyday cooperative behavior: A field experiment,"***Evolution and Human Behavior***, 32, pp. 172–178.

Essler, J. L., S. Marshall-Pescini, & F. Range (2017), "Domestication does not explain the presence of inequity aversion in dogs," ***Current Biology***, 27(12), pp. 1861–1865.

Falk, A., E. Fehr, & U. Fischbacher (2003) "On the nature of fair behavior,"***Economic Inquiry***, 41(1), pp. 20–26.

Fallon, N., C. Roberts, & A, Stancak (2020) "Shared and distinct functional networks for empathy and pain processing: A systematic review and meta-analysis of fMRI studies,"***Social Cognitive and Affective Neuroscience***, 15(7), pp. 709–723.

Farroni, T., G. Csibra, F. Simion, & M. H. Johnson (2002) "Eye contact detection in humans from birth,"***Proceedings of the National Academy of Sciences***, 99(14), pp. 9602–9605.

Farroni, T., M. H. Johnson, E. Menon, L. Zulian, D. Faraguna, & G. Csibra (2005) "Newborns'preference for face-relevant stimuli: Effects of contrast polarity,"***Proceedings of the National Academy of Sciences***, 102(47), pp. 17245–17250.

Fehr, E., H. Bernhard, & B. Rockenbach (2008) "Egalitarianism in young children,"***Nature***, 454(7208), pp. 1079–1083.

Fehr, E., & U. Fischbacher (2004) "Third-party punishment and social norms,"***Evolution and Human Behavior***, 25(2), pp. 63–87.

Fehr, E., & S. Gächter (2000) "Cooperation and punishment in public goods experiments,"***American Economic Review***, 90(4), pp. 980–994.
Fehr, E., & S. Gächter (2002) "Altruistic punishment in humans,"***Nature***, 415, pp. 137–140.
Fehr, R., M. J. Gelfand, & M. Nag, (2010), "The road to forgiveness: a meta-analytic synthesis of its situational and dispositional correlates," ***Psychological Bulletin***, 136(5), pp. 894–914.
Fehr, E., & B. Rockenbach (2003) "Detrimental effects of sanctions on human altruism,"***Nature***, 422(6928), pp. 137–140.
Fehr, E., & S.-V. Renninger (2004) "The Samaritan paradox,"Scientific American Mind, 14(5), pp. 14–21.
Fehr, E., & K. M. Schmidt (1999) "A theory of fairness, competition, and cooperation,"***Quarterly Journal of Economics***, 114(3), pp. 817–868.
Fehr, E., & F. Schneider (2009) "Eyes are on us, but nobody cares: are eye cues relevant for strong reciprocity? "***Proceedings of the Royal Society B: Biological Sciences***, 277 (1686), pp. 1315–1323.
Fehr, D., & M. Sutter (2019) "Gossip and the efficiency of interactions,"Games and Economic Behavior, 113(C), pp. 448–460.
Feinberg, M., R. Willer, & M. Schultz (2014) "Gossip and ostracism promote cooperation in groups,"***Psychological Science***, 25(3), pp. 656–664.
Fenzl, T., &T, Brudermann, T. (2021). Eye cues increase cooperation in the dictator game under physical attendance of a recipient, but not for all. ***Journal of Behavioral and Experimental Economics***, 94, 101748.
Fiddick, L. (2004) "Domains of deontic reasoning: Resolving the discrepancy between the cognitive and moral reasoning literature,"***Quarterly Journal of Experimental Psychology***, Section A, 57(3), pp. 447–474.
Fiddick, L. (2011) "From social exchange to safety: Further explorations of adaptive reasoning"in X. T. Wang (ed) Thus spake evolutionary psychologists, pp. 136–145, Peking University Press.
Fiddick, L., M. V. Spampianto, & J. Graman (2005) "Social contracts and precautions activate different neurological systems: An fMRI investigation of deontic reasoning,"***NeuroImage***, 28(4), pp. 778–786.
Fields, D. (2016) Why we snap: Understanding the rage circuit in your brain, Dutton.
Fisman, R., S. Kariv, & D. Markovits (2007) "Individual preferences for giving,"American Economic Review, 97(5), pp. 1858–1876.
Fonseca, M. A., & K. Peters (2018) "Will any gossip do? Gossip does not need to be perfectly accurate to promote trust,"***Games and Economic Behavior***, 107, pp. 253–281.
Ford, B. Q., J. O. Dmitrieva, D. Heller, Y. Chentsova-Dutton, I. Grossmann, M. Tamir, Y. Uchida, B. Koopmann-Holm, B. A. Floerke, M. Uhrig, T. Bokhan & I. B. Mauss, (2015), "Culture shapes whether the pursuit of happiness predicts higher or lower well-being." ***Journal of Experimental Psychology: General***, 144(6), pp. 1053–1062.
Ford, Q. B., A J. Shallcross, I. B. Mauss, V. A. Floerke & J. Gruber, (2014), "Desperately Seeking Happiness: Valuing Happiness is Associated With Symptoms and Diagnosis of Depression," ***Journal of Social and Clinical Psychology***, 33(10).
Foster, E. K. (2004) "Research on gossip: Taxonomy, methods, and future directions,"Review of General Psychology, 8(2), pp. 78–99.
Francey, D, & R. Bergmüller (2012) "Images of eyes enhance investments in a real-life public good,"***PLoS ONE***, 7(5): e37397.
Frank, R. H. (1988) Passions within reason: The strategic role of emotions, W. W. Norton & Co Inc.
Franklin, B. (1962/c.1791) Autobiography of Benjamin Franklin, McMillan.
Franzen, A., & S. Pointner (2012) "Anonymity in the dictator game revisited,"***Journal of Economic Behavior and Organization***, 81(1), pp. 74–81.

Freak-Poli, R., N. Wagemaker, R. Wang, T. S. Lysen, M. A. Ikram, M. W. Vernooij, C. S. Dintica, M. Vernooij-Dassen, R. J. F. Melish, E. J. Laukka, L. Fratiglionic, W. Xu, & H. Tiemeier, (2022). Loneliness, not social support, is associated with cognitive decline and dementia across two longitudinal population-based cohorts. ***Journal of Alzheimer's Disease***, 85(1), pp. 295–308.

Friedman, M., & R. Friedman (1990) Free to choose: A personal statement, Houghton Mifflin Harcourt.

Fudenberg, D., & P. A. Pathak (2010) "Unobserved punishment supports cooperation,"***Journal of Public Economics***, 94(1–2), pp. 78–86.

Fujii, N. (2009) "The Connected Brain (Tsunagaru nou)" (in Japanese) NTT Publishing

Fujii, N. (2010) "Introduction to Social Brains: what is the 'social brain'? (Sousharu bureinzu nyumon: shakai nou)" (in Japanese) Kodansha

Funk, F., V. McGeer, & M. Gollwitzer (2014) "Get the message: Punishment is satisfying if the transgressor responds to its communicative intent,"***Personality and Social Psychology Bulletin***, 40(8), pp. 986–997.

Fumagalli, E., L. J. Shrum, & T. M. Lowrey (2022) "Consuming in response to loneliness: Bright side and dark side effects,"***Current Opinion in Psychology***, 101329.

Gabay, A. S., J. Radua, M. J. Kempton, & M. A. Mehta (2014) "The ultimatum game and the brain: A meta-analysis of neuroimaging studies,"***Neuroscience & Biobehavioral Reviews***, 47, pp. 549–558.

Gächter, S., B. Herrmann, & C. Thöni (2010) "Culture and cooperation,"***Philosophical Transactions of the Royal Society B***, 365(1553), pp. 2651–2661.

Gächter, S., E. Renner, & M. Sefton (2008) "The long-run benefits of punishment,"***Science***, 322 (5907), pp. 1510–1510.

Gazzaniga, M. S. (2008) Human-The science behind what makes us unique, Ecco.

Gentile, D. A., Sweet, D. M., & He, L. (2020) "Caring for others cares for the self: an experimental test of brief downward social comparison, loving-kindness, and interconnectedness contemplations," ***Journal of Happiness Studies***, 21(3), pp. 765–778.

Gerfo, E. L., A. Gallucci, R. Morese, A. Vergallito, S. Ottone, F. Ponzano,G. Locatelli, F. Bosco, & L. J. R. Lauro (2019) "The role of ventromedial prefrontal cortex and temporo-parietal junction in third-party punishment behavior,"***Neuroimage***, 200, pp. 501–510.

Giedd, J. N. (2015) "The amazing teen brain,"***Scientific American***, 312(6), pp. 32–37.

Gigerenzer, G. (2007) Gut feelings: The intelligence of the unconscious, Viking Adult.

Gigerenzer, G., & K. Hug (1995) "Domain-specific reasoning: Social contracts, cheating, and perspective change,"***Cognition***, 43(2), pp. 127–171.

Gintis, H. (2009) The bounds of reason: Game theory and the unification of the behavioral sciences, Princeton University Press.

Gneezy, U., & A. Rustichini (2000) "Pay enough or don't pay at all,"***The Quarterly Journal of Economics***, 115(3), pp. 791–810.

Goette, L., D. Huffman, & S. Meier (2006) "The impact of group membership on cooperation and norm enforcement: Evidence using random assignment to real social groups,"***American Economic Review***, 96(2), pp. 212–216.

Gollwitzer, M., M. Meder, & M. Schmitt (2011) "What gives victims satisfaction when they seek revenge? "***European Journal of Social Psychology***, 41(3), pp. 364–374.

Gonsalkorale, K., & K. D. Williams (2007) "The KKK won't let me play: Ostracism even by a despised outgroup hurts,"***European Journal of Social Psychology***, 37(6), pp. 1176–1186.

Gordon, A. D., & Schweitzer, M. E. (2024) Gossip, power, and advice: Gossipers are conferred less expert power. ***Journal of Experimental Social Psychology***, 115, 104655.

Gould, S. J., & R. C. Lewontin (1979) "The spandrels of San Marco and the Panglossian paradigm: A critique of the adaptationist programme,"***Proceedings of the Royal Society B, Biological Sciences***, 205(1161), pp. 581–598.

Gray K, & D. M. Wegner, (2008), "The sting of intentional pain," ***Psychological Science***,19(12), pp. 1260–1262.

Greene, J. (2013) Moral tribes: Emotion, reason, and the gap between us and them, Penguin Press.

Giesen, A., & G. Echterhoff (2018) "Do I really feel your pain? Comparing the effects of observed and personal ostracism,"***Personality and Social Psychology Bulletin***, 44(4), pp. 550–561.

Grocke, P., F. Rossano, & M. Tomasello (2015) "Procedural justice in children: Preschoolers accept unequal resource distributions if the procedure provides equal opportunities,"***Journal of Experimental Child Psychology***, 140, pp. 197–210.

Grossmann, T., M. H. Johnson, S. Lloyd-Fox, A. Blasi, F. Deligianni, C. Elwell, & G. Csibra (2008) "Early cortical specialization for face-to-face communication in human infants,"***Proceedings of the Royal Society B: Biological Sciences***, 275(1653), pp. 2803–2811.

Grossmann, Z., & J. J. van der Weele (2017) "Self-image and willful ignorance in social decisions,"***Journal of European Economic Association***, 15(1), pp. 173–217.

Guala, F. (2012) "Reciprocity: Weak or strong? What punishment experiments do (and do not) demonstrate,"***Behavioral and Brain Science***, 35(1), pp. 1–15.

Gürerk, Ö., B. Irlenbusch, & B. Rockenbach (2006) "The competitive advantage of sanctioning institutions,"***Science***, 312(5770), pp. 108–111.

Güth, W., & M. G. Kocher (2014) "More than thirty years of ultimatum bargaining experiments: Motives, variations, and a survey of the recent literature,"***Journal of Economic Behavior & Organization***, 108, pp. 396–409.

Güth, W., R. Schmittberger, & B. Schwarze (1982) "An experimental analysis of ultimatum bargaining,"***Journal of Economic Behavior and Organization***, 3(4), pp. 367–388.

Güth, W., & R. Tietz (1990) "Ultimatum bargaining behavior: A survey and comparison of experimental results,"***Journal of Economic Psychology***, 11(3), pp. 417–449.

Guthrie, S. E. (1993) Faces in the clouds: A new theory of religion, Oxford University Press.

Haidt, J. (2006) The happiness hypothesis: Finding modern truth in ancient wisdom, Basic Books.

Haidt, J. (2012) The righteous mind: Why good people are divided by politics and religion, Pantheon.

Haley, K. J., & D. M. T. Fessler (2005) "Nobody's watching? Subtle cues affect generosity in an anonymous economic game,"***Evolution and Human Behavior***, 26(3), pp. 245–256.

Hamilton, W. D. (1964) "The Genetical Evolution of Social Behaviour, I, II,"***Journal of Theoretical Biology***, 7(1), pp. 1–52.

Hamlin, J. K. (2013) "Failed attempts to help and harm: Intention versus outcome in preverbal infants'social evaluations,"***Cognition***, 128(3), pp. 451–474.

Hamlin J. K., K. Wynn, & P. Bloom (2007) "Social evaluation by preverbal infants,"***Nature***, 450(7169), pp. 557–559.

Hamlin, J. K., K. Wynn, & P. Bloom (2010) "Three-month-olds show a negativity bias in their social evaluation,"***Developmental Science***, 13(6), pp. 923–929.

Hamlin, J. K., K. Wynn, P. Bloom, & N. Mahajan (2011) "How infants and toddlers react to antisocial others,"***Proceedings of the National Academy of Sciences***, 108(50), pp. 19931–19936.

Harbaugh, W. T., U. Mayr, & D. R. Burghart (2007) "Neural responses to taxation and voluntary giving reveal motives for charitable donations,"***Science***, 316(5831), pp. 1622–1625.

Harris, L. T., & S. T. Fiske (2006) "Dehumanizing the lowest of the low: Neuroimaging responses to extreme out-groups,"Psychological Science, 17(10), pp. 847–853.

Harris, P. L., M. Núñez, & C. Brett (2001) "Let's swap: Early understanding of social exchange by British and Nepali children,"***Memory & Cognition***, 29(5), pp. 757–764.

Harris, C., A. Myers, & A. Kaiser, (2023), "The humanizing effect of market interaction," ***Journal of Economic Behavior & Organization***, 205, pp. 489–507.

Hartsough, L. E. S., R. Ginther, & R. Marois (2020) "Distinct affective responses to second- and third-party norm violations,"***Acta Psychologica***, 205, 103060.

Haruno, M., & C. D. Frith (2010) "Activity in the amygdala elicited by unfair divisions predicts social value orientation,"***Nature Neuroscience***, 13(2), pp. 160–161.

Hasegawa, J. & K. Hiraishi (2000) 'The Development of the Mind from the Perspective of Evolutionary Psychology (Shinka shinrigaku kara mita kokoro no hassei)', Chapter 8 in Comparative Cognitive Science of the Mind (Development of Language and Mind) (Kokoro no hikaku ninchi Kagaku (Kotoba to kokoro no hattatsu), edited by Shigeru Watanabe, supervised by Shozo Kojima and Hiroto Katori, Minerva Shobo. (in Japanese), Minerva publishing

Hasegawa, M. (2016) "Human Prosociality from the Perspective of Evolutionary Psychology (Shinka sinrigaku kara mita hito no shakaisei (kyoukan))," ***Japanese Journal of Cognitive Neuroscienc***, 18(3+4) pp. 108–114 (in Japanese).

Haselton, M. G., & D. M. Buss (2000) "Error management theory: A new perspective on biases in cross-sex mind reading,"***Journal of Personality and Social Psychology***, 78(1), pp. 81–91.

Haselton, M. G., G. A. Bryant, A. Wilke, D. A. Frederick, A. Galperin, W. E. Frankenhuis, & T. Moore (2009) "Adaptive rationality: An evolutionary perspective on cognitive bias,"***Social Cognition***, 27(5), pp. 733–763.

Hashimoto, H., N. Mifune, & T. Yamagishi (2014) "To be perceived as altruistic: Strategic considerations that support fair behavior in the dictator game,"***Letters on Evolutionary Behavioral Science***, 5(2), pp. 17–20.

Hauser, O. P., M. A. Nowak, & D. G. Rand (2014) "Punishment does not promote cooperation under exploration dynamics when anti-social punishment is possible,"***Journal of Theoretical Biology***, 360, pp. 163–171.

Heath, C., C. Bell, & E. Sternberg (2001) "Emotional selection in memes: the case of urban legends,"Journal of Personality and Social Psychology, 81(6), pp. 1028–1041.

Hechler, S. F. Funk, & T. Kessler, (2023), "Not revenge, but change is sweet: Experimental evidence of how offender change and punishment play independent roles in victims' sense of justice," British Journal of Social Psychology, 62(2), pp. 1013–1035.

Hein, G., G. Silani, K. Preuschoff, C. D. Batson, & T. Singer (2010) "Neural responses to ingroup and outgroup members' suffering predict individual differences in costly helping,"***Neuron***, 68(1), pp. 149–160.

Henrich, J. (2004) "Animal behaviour (communication arising) : Inequity aversion in capuchins?,"***Nature***, 428(6979), p.139.

Henrich, J. (2017) "The secret of our success: How culture is driving human evolution, domesticating our species, and making us smarter, Princeton University Press.

Henrich, J. (2020) "the WEIRDest people in the world: How the west became psychologically peculiar and particularly prosperous," Allen Lane

Henrich, J., R. Boyd, S. Bowles, C. Camerer, E. Fehr, H. Gintis, & R. McElreath (2001) "In search of homo economicus: Behavioral experiments in 15 small-scale societies,"***American Economic Reviews***, 91(2), pp. 73–78.

Henrich, J., R. Boyd, S. Bowles, C. F. Camerer, E. Fehr, H. Gintis, & R. McElreath (2004) "Overview and synthesis," in J. Henrich, R. Boyd, S. Bowles, C. F. Camerer, E. Fehr, & H. Gintis (Eds.), Foundations of human sociality: Economic experiments and ethnographic evidence from fifteen small-scale societies, Oxford University Press.

Henrich, J., J. Ensminger, R. McElreath, A. Barr, C. Barrett, A. Bolyanatz, J. C. Cardenas, M. Gurven, E. Gwako, N. Henrich, C. Lesorogol, F. Marlowe, D. Tracer, & J. Ziker (2010a) "Markets, religion, community size, and the evolution of fairness and punishment," ***Science***, 327(5972), pp. 1480–1484.

Henrich, J., S. J. Heine, & A. Norenzayan (2010b) "The weirdest people in the world?,"***Behavioral and Brain Sciences***, 33(2–3), pp. 61–83.

Henrich, J., R. McElreath, A. Barr, J. Ensminger, C. Barrett, A. Bolyanatz, J. C. Cardenas, M. Gurven, E. Gwako, N. Henrich, C. Lesorogol, F. Marlowe, D. Tracer, & J. Ziker (2006) "Costly punishment across human societies,"***Science***, 312(5781), pp. 1767–1770.

Hepach, R., A. Vaish, and M. Tomasello (2012) "Young children are intrinsically motivated to see others helped," ***Psychological Science***, 23(9), pp. 967–972.

Hernandez-Lallement, J., A. T. Attah, E. Soyman, C. M. Pinhal, V. Gazzola, & C. Keysers (2020) "Harm to others acts as a negative reinforcer in rats,"***Current Biology***, 30(6), pp. 949–961.

Herrmann, B., C. Thöni, & S. Gächter (2008) "Antisocial punishment across societies,"***Science***, 319(5868), pp. 1362–1367.

Hess, N. H., & E. H. Hagen (2006) "Sex differences in indirect aggression: Psychological evidence from young adults,"***Evolution and Human Behavior***, 27(3), pp. 231–245.

Hoeft, L., & W. Mill (2017) "Selfish punishers: An experimental investigation of designated punishment behavior in public goods,"***Economics Letters***, 157, pp. 41–44.

Hoes, E., B. Aitken, J. Zhang, T. Gackowski & M. Wojcieszak (2024), "Prominent misinformation interventions reduce misperceptions but increase scepticism," Nature Human Behaviour.

Hoffman, E., K. McCabe, K. Shachat, & V. Smith (1994) "Preferences, property rights and anonymity in bargaining games,"***Games and Economic Behavior***, 7(3), pp. 346–380.

Hofmann, W., D. C. Wisneski, M. J. Brandt, & L. J. Skitka (2014) "Morality in everyday life," ***Science***, 345(6202), pp. 1340–1343.

Holt-Lunstad, J., T. B. Smith, M. Baker, T. Harris, & D. Stephenson (2015) "Loneliness and social isolation as risk factors for mortality: a meta-analytic review,"Perspectives on Psychological Science, 10(2), pp. 227–237.

Hood, B. M., C. N. Macrae, V. Cole-Davies, & M. Dias (2003) "Eye remember you: The effects of gaze direction on face recognition in children and adults," *Developmental Science*, 6(1), pp. 67–71.

Hui, B. P. H., Ng, J. C. K., Berzaghi, E., Cunningham-Amos, L. A., & Kogan, A (2020) "Rewards of kindness? A meta-analysis of the link between prosociality and well-being," ***Psychological Bulletin***, 146(12), 1084–1116.

Ingram, G. P. D., & J. M. Bering (2010) "Children's tattling: The reporting of everyday norm violations in preschool settings,"Child Development, 81(3), pp. 945–957.

Izuma K., D. N. Saito, & N. Sadato (2008) "Processing of social and monetary rewards in the human striatum,"***Neuron***, 58(2), pp. 284–294.

Jacobs, J. (1994). Systems of survival: A dialogue on the moral foundations of commerce and politics. Vintage.

James, S. M. (2010) An introduction to evolutionary ethics, Wiley-Blackwell.)

Jensen, K., J. Call, & M. Tomasello (2007a) "Chimpanzees are rational maximizers in an ultimatum game,"***Science***, 318(5847), pp. 107–109.

Jensen, K., J. Call, & M. Tomasello (2007b) "Chimpanzees are vengeful but not spiteful,"***Proceedings of the National Academy of Sciences***, 104(32), pp. 13046–13050.

Jensen, K., & M. Tomasello (2019) "Punishment,"Encyclopedia of Animal Behavior (Second Edition), Academic Press, pp. 214–219.

Jiang, X., N. Zhang, X. Sun, Z. Liu, & Y. L. Wang, (2024) "Being pressed for time leads to treating others as things: Exploring the relationships among time scarcity, agentic and communal orientation and objectification,"British Journal of Social Psychology.

Jiménez-Zafra, S. M., A. J. Sáez-Castillo, A. Conde-Sánchez & M. T. Martín-Valdivia, (2021), "How do sentiments affect virality on Twitter?", ***Royal Society Open Science***, 8: 201756

Jordan, J. J., K. McAuliffe, & F. Warneken (2014) "Development of in-group favoritism in children's third-party punishment of selfishness,"***Proceedings of the National Academy of Sciences***, 111(35), pp. 12710–12715.

Jordan, J. J., M. Hoffman, P. Bloom, & D. G. Rand (2016a) "Third-party punishment as a costly signal of trustworthiness,"***Nature***, 530(7591), pp. 473–476.

Jordan, J. J., K. McAuliffe, & D. G. Rand (2016b) "The effects of endowment size and strategy method on third party punishment,"***Experimental Economics***, 19(4), pp. 741–763.

Jordan, J. J., M. Hoffman, M. A. Nowak, & D. G. Rand, (2016c) "Uncalculating cooperation is used to signal trustworthiness,"***Proceedings of the National Academy of Sciences***, 113(31), pp. 8658–8663.

Kahneman, D. (2011) Thinking fast and slow, Farrar, Straus and Giroux.

Kaiser I, K. Jensen, J. Call, M. Tomasello, 2012 "Theft in an ultimatum game: chimpanzees and bonobos are insensitive to unfairness," ***Biology Letters***, 8(6), pp. 942–945.

Kamei, K. (2018) "The role of visibility on third party punishment actions for the enforcement of social norms,"***Economic Letters***, 171, pp. 193–197.

Kamei, K. (2020) "Group size effect and over-punishment in the case of third party enforcement of social norms,"***Journal of Economic Behavior & Organization***, 175, pp. 395–412.

Kanakogi, Y., Y. Inoue, G. Matsuda, D. Butler, K. Hiraki, & M. Myowa-Yamakoshi (2017) "Preverbal infants affirm third-party interventions that protect victims from aggressors,"***Nature Human Behaviour***, 1: 0037.

Kanakogi, Y., M. Miyazaki, H. Takahashi, H. Yamamoto, T. Kobayashi, & K. Hiraki (2022) "Third-party punishment by preverbal infants." ***Natture Human Behaviour***, 6, pp. 1234–1242.

Karagözolu, E., & U. B. Urhan (2017) "The effect of stake size in experimental bargaining and distribution games: A survey,"***Group Decision and Negotiation***, 26(2), pp. 285–325.

Kawai, N., A. Nakagami, M. Yasue, H. Koda, & N. Ichinohe (2019) "Common marmosets (Callithrix jacchus) evaluate third-party social interactions of human actors but Japanese monkeys (Macaca fuscata) do not,"***Journal of Comparative Psychology***, 133(4), pp. 488–495.

Kawai, N., M. Yasue, T. Banno, & N. Ichinohe (2014) "Marmoset monkeys evaluate third-party reciprocity,"***Biology Letters***, 10(5).

Kawamura, Y., & T. Kusumi (2017) "The norm-dependent effect of watching eyes on donation,"***Evolution and Human Behavior***, 38(5), pp. 659–666.

Katzir, M., S. Cohen, & E. Halali, (2021) "Is it all about appearance? Limited cognitive control and information advantage reveal self-serving reciprocity,"***Journal of Experimental Social Psychology***, 97, 104192.

Kelsey, C., A. Vaish, & T. Grossmann, (2018) "Eyes, more than other facial features, enhance real-world donation behavior,"***Human Nature : An Interdisciplinary Biosocial Perspective***, 29(4), pp. 390–401.

Kennedy, D., & C. Norman (2005) "What we don't know,"***Science***, 309(5731), pp. 75–93.

Kenrick, D. T. (2011) Sex, Murder, and the Meaning of Life: A psychologist investigates how evolution, cognition, and complexity are revolutionizing our view of human nature, Basic Books.

Kenrick, D. T., & V. Griskevicius (2013) The rational animal: How evolution made us smarter than we think, Basic Books.

Kenrick, D. T., V. Griskevicius, J. M. Sundie, N. P. Li, Y. J. Li, & S. L. Neuberg (2009) "Deep Rationality: The Evolutionary Economics of Decision Making,"***Social Cognition***, 27(5), pp. 764–785.

Kinzler, K. D., & K. Shutts (2008) "Memory for mean over nice the influence of threat on children's face memory,"Cognition, 107(2), pp. 775–783.

Kiyonari, T., S. Tanida, & T. Yamagishi (2000) "Social exchange and reciprocity: Confusion or a heuristic?,"***Evolution and Human Behavior***, 21(6), pp. 411–427.

Kniffin, K. M., & D. S. Wilson (2005) "Utilities of gossip across organizational levels,"***Human Nature***, 16(3), pp. 278–292.

Knoch, D., A. Pascual-Leone, K. Meyer, V. Treyer, & E. Fehr (2006) "Diminishing reciprocal fairness by disrupting the right prefrontal cortex,"***Science***, 314(5800), pp. 829–832.

Kohouta, S., S. Kruikemeier, B. N. Bakker, (2023), "May I have your Attention, please? An eye tracking study on emotional social media comments," ***Computers in Human Behavior***, 139, 107495.

Koornneef, E. J., A. Dariel, I. Elbarazi, A. R. Alsuwaidi, P. B. M. Robben,, & N. Nikiforakis, (2018) "Surveillance cues do not enhance altruistic behavior among anonymous strangers in the field,"***PLoS ONE*** 13: e0197959.

Knutson, B. (2004) "Sweet revenge? "***Science***, 305(5688), pp. 1246–1247.

Kobayashi, H. (2019) "The Moai's eye (Moai no shiro me: me to kokoro no kininaru kankei)," University of Tokyo press (in Japanese).

Kobayashi, H. (2022) " Flightless lizards: Welcome to the forest of science," (Tobanai tokage: Youkoso saiensu no mori he) University of Tokyo press (in Japanese).

Kobayashi, K. (2021) "A role of anger from the perspective of game theory: A responder in an ultimatum game as an example," ***Cognitive Studies: Bulletin of the Japanese Cognitive Science Society***, 28(3), pp. 445–457. (in Japanese)

Kobayashi, K (2024) "When I catch your eye, I can't help but straighten up (Anata to me ga auto chanto shitesimau watashi" in Science of the Mechanism of "Fit": The Psychology of "You" and "Me", Sakaguchi, Y. & K. Tomida(eds), Minerva Shobo, (in Japanese)

Kobayashi, K. (mimeo) One-shot games and error management theory; ~ why do we behave safely when we meet people for the first time? (in Japanese)

Kobayashi, K., & T. Motonishi (mimeo) "Does the Cheater Detection Shield Consumers from Greedy Financial Product Salespeople?"

Kodandaramaiah, U., A. Vallin, & C. Wiklund (2009) "Fixed eyespot display in a butterfly thwarts attacking birds,"***Animal Behaviour***, 77(6), pp. 1415–1419.

Korenok, O., E. L. Millner, & L. Razzolini (2014) "Taking, giving, and impure altruism in dictator games,"***Experimental Economics***, 17(3), pp. 488–500.

Krupka, E. L., & R. T. Croson (2016) "The differential impact of social norms cues on charitable contributions," ***Journal of Economic Behavior & Organization***, 128, pp. 149–158.

Kurzban, R. (2011) Why everyone (else) is a hypocrite: Evolution and the modular mind, Princeton University Press.

Krys, K., Kostoula, O., van Tilburg, W. A. P., Mosca, O., Lee, J. H., Maricchiolo, F., Kosiarczyk, A., Kocimska-Bortnowska, A., Torres, C., Hitokoto, H., Liew, K., Bond, M. H., Lun, V. M.-C., Vignoles, V. L., Zelenski, J. M., Haas, B. W., Park, J., Vauclair, C.-M., Kwiatkowska, A., … Uchida, Y. (2024), "Happiness Maximization Is a WEIRD Way of Living," Perspectives on Psychological Science, 0(0).

Lamba, S., & R. Mace, (2010) "People recognize when they are really anonymous in an economic game,"***Evolution and Human Behavior***, 31(4), pp. 271–278.

Lamm, C., J. Decety, & T. Singer (2011) "Meta-analytic evidence for common and distinct neural networks associated with directly experienced pain and empathy for pain,"***Neuroimage***, 54(3), pp. 2492–2502.

Langford, D. J., Crager, S. E., Shehzad, Z., Smith, S. B., Sotocinal, S. G., Levenstadt, J. S., M. L. Chanda, D. J. Levitin, & J. S. Mogil (2006) "Social modulation of pain as evidence for empathy in mice,"***Science***, 312(5782), pp. 1967–1970.

Ledyard, J. O. (1995) "Public goods: A survey of experimental research,"in J. Kagel, & A. Roth (Eds.), The handbook of experimental economics, Chap. 2, pp. 111–194, Princeton University Press.

Lee, Y., & F. Warneken (2020) "Children's evaluations of third-party responses to unfairness: Children prefer helping over punishment,"***Cognition***, 205, 104374.

Legate, N., C. R. DeHaan, N. Weinstein, & R. M. Ryan (2013) "Hurting you hurts me too: The psychological costs of complying with ostracism,"***Psychological Science***, 24(4), pp. 583–588.

Leimgruber, K. L., A. G. Rosati, & L. R. Santos (2016) "Capuchin monkeys punish those who have more,"***Evolution and Human Behavior***, 37(3), pp. 236–244.

Lennon, P., R. Grant, & V. T. Montrose (2017) "Stylized and photographic eye images do not increase charitable donations in a field experiment,"***Letters on Evolutionary Behavioral Science***, 8, pp. 28–31.

Levitt, S. D., & S. J. Dubner (2009) "SuperFreaconomics: Global cooling, patriotic prostitutes, and why suicide bombers should buy life insurance,"William Morrow.

Levitt, S. D., & J. A. List (2008) "Homo Economicus Evolves,"***Science***, 319(5865), pp. 909–910.

Li, X., M. Jusup, Z. Wang, H. Li, L. Shi, B. Podobnik, H. E. Stanley, S. Halvin, & S. Boccaletti (2018) "Punishment diminishes the benefits of network reciprocity in social dilemma experiments,"***Proceedings of the National Academy of Sciences***, 115(1), pp. 30–35.

Li, Y., & N., Mifune, (2023) "Punishment in the public goods game is evaluated negatively irrespective of non-cooperators' motivation," ***Frontiers in Psychology***, 14, 1198797.

Li, M., C. Shangguan, H. Shi, & J. Lu, (2021) "Watching eyes triggers third party punishment the role of emotion within the eyes,"***Fontiers in Psychology***, 12, 681664

Lian, H., J. K. Li, J. Pan, C. Du, & Q. Zhao, (2022) "Are gossipers looked down upon? A norm-based perspective on the relation between gossip and gossiper status,"***The Journal of Applied Psychology***,

Lieberman, D. E. (2013a) "The story of the human body: Evolution, health, and disease, "Pantheon.

Lieberman, M. D. (2013b) "Why our brains are wired to connect, Crown,"

Linden, D. J. (2016) Touch: The science of hand, heart, and mind, Penguin Books.

Lim, K. T. K, & R. Yu (2015) "Aging and wisdom: Age-related changes in economic and social decision making,"***Frontiers in Aging Neuroscience***, 7(120).

List, J. A. (2007) "On the interpretation of giving in dictator games,"***Journal of Political Economy***, 115(3), pp. 482–493.

List, J. A., R. P. Berrens, A. K. Bohara, & J. Kerkvliet (2004) "Examining the role of social isolation on stated preferences,"***American Economic Review***, 94(3), pp. 741–752.

Lucas, R. E., & P. S. Dyrenforth (2006) "Does the existence of social relationships matter for subjective well-being?,"In K. D. Vohs, & E. J. Finkel (Eds.), Self and relationships: Connecting intrapersonal and interpersonal processes (pp. 254–273).

Lucca, K., Yuen, F., Wang, Y., Alessandroni, N., Allison, O., Alvarez, M., Axelsson, E. L., Baumer, J., Baumgartner, H. A., Bertels, J., Bhavsar, M., Byers-Heinlein, K., Capelier-Mourguy, A., Chijiiwa, H., Chin, C. S., Christner, N., Cirelli, L. K., Corbit, J., Daum, M. M., Doan, T., Dresel M., Exner A., Fei W., Forbes S.H., Franchin L., Frank M.C., Geraci A., Giraud M., Gornik M.E., Wiesmann C.G., Grossmann T., Hadley I.M., Havron N., Henderson A.M.E., Matzner E.H., Immel B.A., Jankiewicz G., Jędryczka W., Kanakogi Y., Kominsky J.F., Lew-Williams C., Liberman Z., Liu L., Liu Y., Loeffler M.T., Martin A., Mayor J., Meng X., Misiak M., Moreau D., Nencheva M.L., Oña L.S., Otálora Y., Paulus M., Pepe B., Pickron C.B., Powell L.J., Proft M., Quinn A.A., Rakoczy H., Reschke P.J., Roth-Hanania R., Rothmaler K., Schlegelmilch K., Schlingloff-Nemecz L., Schmuckler M.A., Schuwerk T., Seehagen S., Şen H.H., Shainy M.R., Silvestri V., Soderstrom M., Sommerville J., Song H.J., Sorokowski P., Stutz S.E., Su Y., Taborda-Osorio H., Tan A.W.M., Tatone D., Taylor-Partridge T., Tsang C.K.A., Urbanek A., Uzefovsky F., Visser I., Wertz A.E., Williams M., Wolsey K., Wong T.T., Woodward A.M., Wu Y., Zeng Z., Zimmer L., Hamlin, J. K (2025) "Infants' Social Evaluation of Helpers and Hinderers: A Large-Scale, Multi-Lab, Coordinated Replication Study," ***Developmental Science***, 28(1), e13581.

Ma, N., N. Li, X.-S. He, D.-L. Sun, X. Zhang, & D.-R. Zhang (2012) "Rejection of unfair offers can be driven by negative emotions. evidence from modified ultimatum games with anonymity," ***PLoS ONE***, 7(6): e39619.

Maestripieri, D (2012) Games primates play: An undercover investigation of the evolution and economics of human relationships, Basic Book.

Manesi, Z., & T. Pollet, (2017) "No support for the watching eyes effect across three" lost letter" field experiments,"***Letters on Evolutionary Behavioral Science***, 8(1), pp. 12–15.

Manesi, Z., P. A. M. Van Lange, & T. V. Pollet, (2016) "Eyes wide open: Only eyes that pay attention promote prosocial behavior,"***Evolutionary Psychology***, 14(2), pp. 1–15.

Margoni, F., & L. Surian (2018) "Infants'evaluation of prosocial and antisocial agents: A meta-analysis,"***Developmental Psychology***, 54(8), pp. 1445–1455.

Marlowe, F. W., (2009) "Hadza cooperation: Second-party punishment, yes; Third-party punishment, no," ***Human Nature***, 20(4), pp. 417–430.

Marlowe, F. W., J. C. Berbesque, A. Barr, C. Barrett, A, Bolyanatz, J. C. Cardenas,... & N. Henrich, (2008) "More 'altruistic'punishment in larger societies," ***Proceedings of the Royal Society B: Biological Sciences***, 275(1634), pp. 587–592.

Marshall, A. (1920) Principles of economics: 8th edition, Macmillan and Co. (first edition 1890)

Mas, A., & E. Moretti (2009) "Peers at work,"***American Economic Review***, 99(1), pp. 112–45.

Mas-Colell, A., M. D. Whinston, & J. R. Green (1995) Microeconomic theory, Oxford University Press.

Mason, M. F., B. M. Hood, & C. N. MacRae, (2004) "Look into my eyes: Gaze direction and person memory, ***Memory***, 12, pp. 637–643.

Masserman, J. H., S. Wechkin, & W. Terris (1964) ""Altruistic" behavior in rhesus monkeys,"***American Journal of Psychiatry***, 121(6), pp. 584–585.

Masten, C. L., N. I. Eisenberger, J. H. Pfeifer, & M. Dapretto (2010) "Witnessing peer rejection during early adolescence: Neural correlates of empathy for experiences of social exclusion,"***Social Neuroscience***, 5(5–6), pp. 496–507.

Masten, C. L., S. A. Morell, & N. I. Eisenberge (2011) "An fMRI investigation of empathy for'social pain'and subsequent prosocial behavior,"***Neuroimage***, 55(1), pp. 381–388.
Matsugasaki, K., W. Tsukamoto, & Y. Ohtsubo (2015) "Two failed replications of the watching eyes effect,"Letters on Evolutionary Behavioral Science, 6(2), pp. 17–20.
Matsumoto, Y., T. Yamagishi, Y. Li, & T. Kiyonari (2016) " Prosocial behavior increases with age across five economic games,"***PloS ONE***, 11(7): e0158671.
Mauss, I. B., M. Tamir, C. L. Anderson & N. S. Savino (2011), "Can seeking happiness make people unhappy? Paradoxical effects of valuing happiness, ***Emotion***, 11(4), pp. 807–815.
McAuliffe, K., J. Jordan, & F. Warneken (2015a) "Costly third-party punishment in young children,"***Cognition***, 134, pp. 1–10.
McAuliffe, K., L. W. Linda, K. Leimgruber, R. Spaulding, P. R. Blake, & L. R. Santos (2015b) "Capuchin monkeys, Cebus apella, show no evidence for inequity aversion in a costly choice task,"***Animal Behaviour***, 103, pp. 65–74.
McAuliffe, K., & Y. Dunham (2021) "Children favor punishment over restoration,"***Developmental Science***, e13093.
McGetrick, J., & R. Friederike (2018) "Inequity aversion in dogs: a review,"***Learning & Behavior***, 46(4), pp. 479–500.
Mealey, L., C. Daood, & M. Krage (1996) "Enhanced memory for faces of cheaters,"***Ethology and Sociobiology***, 17(2), pp. 119–128.
Mehl, B., & A. Buchner (2008) "No enhanced memory for faces of cheaters,"***Evolution and Human Behavior***, 29(1), pp. 35–41.
Mesoudi, A., A. Whiten, & R. Dunbar (2006) "A bias for social information in human cultural transmission,"***British Journal of Psychology***, 97(3), pp. 405–423.
Meyer, M. L., C. L. Masten, Y. Ma, C. Wang, Z. Shi, N. I. Eisenberger, & S. Han (2013) "Empathy for the social suffering of friends and strangers recruits distinct patterns of brain activation,"***Social Cognitive and Affective Neuroscience***, 8(4), pp. 446–454.
Mifune, N., H. Hashimoto, & T. Yamagishi (2010) "Atruism toward in-group members as a reputation mechanism,"***Evolution and Human Behavior***, 31(2), pp. 109–117.
Milinski, M., D. Semmann, & H. J. Krambeck (2002) "Reputation helps solve the'tragedy of the commons'."***Nature***, 415 (6870), pp. 424–426.
Mitchell, J. P., C N. Macrae, & M. R. Banaji (2006) "Dissociable medial prefrontal contributions to judgments of similar and dissimilar others,"***Neuron***, 50(4), pp. 655–663.
Miyatake, S., & M. Higuchi (2017) "Does religious priming increase the prosocial behaviour of a Japanese sample in an anonymous economic game?" ***Asian Journal of Social Psychology***, 20(1), pp. 54–59.
Moffett, M. W. (2010) Adventures among ants: a global safari with a cast of trillions, Univ of California Press.
Molho, C., J. M. Tybur, P. A. M. Van Lange, & D. Balliet (2020) "Direct and indirect punishment of norm violations in daily life,"***Nature Communications***, 11: 3432.
Molleman, L., F. Kölle, C. Starmer, & S. Gächter, 2019, "People prefer coordinated punishment in cooperative interactions, ***Nature Human Behavior***, 3(11), pp. 1145–1153.
Molnar, A., S. J. Chaudhry & G. Loewenstein (2023) ""It's not about the money. It's about sending a message!" Avengers want offenders to understand the reason for revenge," ***Organizational Behavior and Human Decision Processes***, 174, 104207.
Morewedge, C. K., T. Krishnamurti, & D. Ariely (2014) "Focused on fairness: Alcohol intoxication increases the costly rejection of inequitable rewards," ***Journal of Experimental Social Psychology***, 50, pp. 15–20.
Morimoto, R. & S. Tanaka (2016) "An introduction to philosophy of bilollgy (Seibutsu gaku no tetsugaku nyuumon)," Keiso shobo (in Japanese).
Motterlini, M. (2008) "Emotional economy. What lies behind our daily accounts,"(Keizai ha kanzyou de ugoku) (in Japanese) Kinokuniya. (Motterlini, M. (2006) Economia emotiva. Che cosa si nasconde dietro i nostri conti quotidiani i, BUR Biblioteca Univ. Rizzoli.)

Mulder, L. B., E. Van Dijk, D. De Cremer, & H. A. Wilke (2006) "Undermining trust and cooperation: The paradox of sanctioning systems in social dilemmas,"***Journal of Experimental Social Psychology***, 42(2), pp. 147–162.

Murthy, V. H. (2020) "Together: The Healing Power of Human Connection in a Sometimes Lonely World," Harper

Naclerio, M. E., L. Lazar, E. A. Hornstein, & N. I. Eisenberger (2024) "The well-being paradox: Comparing prosocial and self-kindness interventions for mental health benefits, Emotion", Advance online publication.

Nelson, C. A., 3rd, N. A. Fox, & C. H. Zeanah, Jr (2013) "Anguish of the abandoned child,"***Scientific American***, 308(4), pp. 62–67.

Nesse, R. M. (2005) "Natural selection and the regulation of defenses: A signal detection analysis of the smoke detector principle,"***Evolution and Human Behavior***, 26(1), pp. 88–105.

Nettle,D., Z. Harper, A. Kidson, R. Stone, I. S. Penton-Voak, & M. Bateson (2013) "The watching eyes effect in the Dictator game: It's not how much you give, it's being seen to give something,"***Evolution & Human Behavior***, 34(1), pp. 35–40.

Nettle, D., K. Nott, & M. Bateson (2012) "'Cycle thieves, we are watching you': Impact of a simple signage intervention against bicycle theft,"***PLoS ONE***, 7(12): e51738.

Nihonsugi, T., A. Ihara, & M. Haruno (2015) "Selective increase of intention-based economic decisions by noninvasive brain stimulation to the dorsolateral prefrontal cortex,"***Journal of Neuroscience***, 35(8), pp. 3412–3419.

Nikiforakis, N., C. N. Noussair, & T. Wilkening, (2012) "Normative conflict and feuds: The limits of self-enforcement,"***Journal of Public Economics***, 96(9–10), pp. 797–807.

Nisbett, R. E. (2004) The geography of thought: How Asians and Westerners think differently... and why, Simon and Schuster (Free Press).

Nisbett, R. E., & D. Cohen (1996) Culture of honor: The psychology of violence in the South, Routledge.

Northover, S. B., W. C. Pedersen, A. B. Cohen, & P. W. Andrews (2017a) "Artificial surveillance cues do not increase generosity: Two meta-analyses,"***Evolution and Human Behavior***, 38(1), pp. 144–153.

Northover, S. B., W. C. Pedersen, A. B. Cohen, & P. W. Andrews (2017b) "Effect of artificial surveillance cues on reported moral judgment: Experimental failures to replicate and two meta-analyses,"***Evolution and Human Behavior***, 38(5), pp. 561–571.

Nowak, M. A. (2006) "Five rules for the evolution of cooperation,"***Science***, 314(5805), pp. 1560–1563.

Nowak, M. A. (2012) "Why we help,"Scientific American, July, pp. 34–39.

Nowak, M. A., & K. Sigmund (1998) "Evolution of indirect reciprocity by image scoring,"***Nature***, 393(6685), pp. 573–577.

Oberliessen, L., J. Hernandez-Lallement, S. Schäble, M. van Wingerden, M. Seinstra, & T. Kalenscher, (2016) "Inequity aversion in rats, Rattus norvegicus," ***Animal Behaviour***, 115, pp. 157–166.

Oberliessen, L., & T. Kalenscher, (2019), "Social and non-social mechanisms of inequity aversion in non-human animals," ***Frontiers in Behavioral Neuroscience***, 13, 133.

O'connor, C. (2019) The Origins of Unfairness: Social Categories and Cultural Evolution, Oxford Univ Pr on Demand.

Oda, R. (1997) "Biased face recognition in the prisoner's dilemma game," ***Evolution and Human Behavior***, 18(5), pp. 309–315.

Oda, R. (2019) "Is the watching-eye effect a fluke?,"***Letters on Evolutionary Behavioral Science***, 10(1), pp4–6.

Oda, R., K. Hiraishi, & A. Matsumoto-Oda (2006) "Does an altruist-detection cognitive mechanism function independently of a cheater-detection cognitive mechanism? Studies using Wason selection tasks,"***Evolution and Human Behavior***, 27(5), pp. 366–380.

Oda, R., & R. Ichihashi (2016) "The watching eyes effect on charitable donation is boosted by fewer people in the vicinity,"***Letters on Evolutionary Behavioral Science***, 7(2), pp. 9–12.

Oda, R., Y. Kato, & K. Hiraishi (2015) "The watching-eye effect on prosocial lying,"Evolutionary Psychology, 13(3), pp. 1–5.

Oda, R., & S. Nakajima (2010) "Biased face recognition in the faith game,"***Evolution and Human Behavior***, 31(2), pp. 118–122.

Oda, R., Y. Niwa, A. Honma, & K. Hiraishi (2011) "An eye-like painting enhances the expectation of a good reputation,"***Evolution and Human Behavior***, 32(3), pp. 166–171.

Ogawa, K., T. Kawamura, & K. Matsushita (2020) "Effects of cognitive ability and age on giving in dictator game experiments,"***Research in Economics***, 74(4), pp. 323–335.

Ohtsubo, Y., F. Masuda, E. Watanabe, & A. Masuchi (2010) "Dishonesty invites costly third-party punishment,"***Evolution and Human Behavior***, 31(4), pp. 259–364.

Otsubo, K., Y. Masuda, & H. Yamaguchi (2023) ""Watching Eyes" Do Not Strengthen the Behavioral Intention of Donating Blood: A High-Powered Pre-registered Replication Study,"***Letters on Evolutionary Behavioral Science***, 14(1), pp. 26–31.

Ohtsuki (2014) "Biology of cooperation and punishment (Kyouryoku to batsu no seibutugaku," Iwanami shoten (in Japanese).

Oosterbeek, H., R. Sloof, & G. van de Kuilen (2004) "Cultural differences in ultimatum game experiments: Evidence from a Meta-Analysis,"***Experimental Economics***, 7(2), pp. 171–188.

Ostrom, E. (1990) Governing the commons: The evolution of institutions for collective action, Cambridge University Press.

Panagopoulos, C. (2014) "Watchful eyes: implicit observability cues and voting,"Evolution and Human Behavior, 35(4), pp. 279–284.

Panagopoulos, C., S. van der Linden (2017) "The feeling of being watched: Do eye cues elicit negative affect?,"***North American Journal of Psychology***, 19(1), pp. 113–121.

Pareto, V. (2014) Manual of political economy: A variorum translation and critical edition (Original title: Manuale di Economia Politica (1906)), reprint, A. Montesano et al. (eds), Oxford University Press.

Peng, X., Y. Li, P. Wang, L. Mo, & Q. Chen (2015) "The ugly truth: negative gossip about celebrities and positive gossip about self entertain people in different ways,"***Social Neuroscience***, 10(3), pp. 320–336.

Petkanopoulou, K., R. Rodríguez-Bailón, G. B. Willis & G. A. van Kleef (2019) "Powerless people don't yell but tell: The effects of social power on direct and indirect expression of anger," ***European Journal of Social Psychology***, 49(3), pp. 533–547.

Pfattheicher, S. & J. Keller (2015) "The watching eyes phenomenon; the role of a sense of being seen and public self-awareness," ***European Journal of Social Psychology***, pp. 560–566.

Pfattheicher, S., C. Strauch, S. Diefenbacher, & R. Schnuerch (2018) "A field study on watching eyes and hand hygiene compliance in a public restroom,"***Journal of Applied Social Psychology***, 48(4), pp. 188–194.

Pillutla, M. M., & J. K. Murnighan (1996) "Unfairness, anger, and spite: Emotional rejections of ultimatum offers,"***Organizational Behavior and Human decision Processes***, 68(3), pp. 208–224.

Pinker, S. (2002) The blank state: The modern denial of human nature, Viking.

Poundstone, W. (2010) Priceless: The myth of fair value (and How to take advantage of it), Hill and Wang

Powell, K. L., G. Robert, & D. Nettle (2012) "Eye images increase charitable donations: Evidence from an opportunistic field experiment in a supermarket,"***Ethology***, 118(11), pp. 1096–1101.

Power, M. L., & J. Schulkin (2009) The evolution of obesity, Johns Hopkins University Press.

Proctor, D., R. A. Williamson, F. B. de Waal, & S. F. Brosnan (2013) "Chimpanzees play the ultimatum game,"***Proceedings of the National Academy of Sciences***, 110(6), pp. 2070–2075.

Purzycki, B. G., C. Apicella, Q. D. Atkinson, E. Cohen, R. A. McNamara, A. K. Willard,... & J. Henrich (2016) "Moralistic gods, supernatural punishment and the expansion of human sociality," ***Nature***, 530(7590), pp. 327–330.

Radford, C., J. W. McNutt, T, Rogers, B. Maslen, & N. Jordan (2020) "Artificial eyespots on cattle reduce predation by large carnivores,"***Communications Biology***, 3(1), pp. 1–8.

Rai T. S. (2022) "Material Benefits Crowd Out Moralistic Punishment,"***Psychological Science***, 33(5), pp. 789–797.

Raihani, N. (2021) "Social Instinct: How Cooperation Shaped the World," St Martins Pr.

Raihani, N. J. & R. Bshary (2015a) "The reputation of punishers,"Trends in Ecology & Evolution, 30(2), pp. 98–103.

Raihani, N. J., & R. Bshary (2015b) "Third-party punishers are rewarded, but third-party helpers even more so,"***Evolution***, 69(4), pp. 993–1003.

Rakoczy, H., F. Warneken, & M. Tomasello (2008) "The sources of normativity: young children's awareness of the normative structure of games,"***Developmental psychology***, 44(3), pp. 875–881.

Rakoczy, H., N. Brosche, F. Warneken, & M. Tomasello (2009) "Young children's understanding of the context-relativity of normative rules in conventional games,"***British Journal of Developmental Psychology***, 27(2), pp. 445–456.

Rand, D. G. (2016) "Cooperation, fast and slow: Meta-analytic evidence for a theory of social heuristics and self-interested deliberation,"***Psychological Science***, 27(9), pp. 1192–1206.

Rand, D. G., A. Dreber, T. Ellingsen, D. Fudenberg, & M. A. Nowak (2009a) "Positive interactions promote public cooperation,"***Science***, 325(5945), pp. 1272–1275.

Rand, D. G., H. Ohtsuki, & M. A. Nowak (2009b) "Direct reciprocity with costly punishment: Generous tit-for-tat prevails,"***Journal of Theoretical Biology***, 256(1), pp. 45–57.

Rand, D. G., J. D. Greene, & M. A. Nowak (2012) "Spontaneous giving and calculated greed,"***Nature***, 489(7416), pp. 427–430.

Rand, D. G., A. Peysakhovich, G. T. Kraft-Todd, G. E. Newman, O. Wurzbacher, M. A. Nowak, & J. D. Greene (2014) "Social heuristics shape intuitive cooperation,"***Nature Communications***, 5(1), pp. 1–12.

Range, F., L. Horn, Z. Viranyi, & L. Huber (2009) "The absence of reword induces inequity aversion in dogs,"***Proceedings of the National Academy of Sciences***, 106(1), pp. 340–345.

Range, F., K. Leitner, & Z. Virányi (2012) "The influence of the relationship and motivation on inequity aversion in dogs," ***Social Justice Research***, 25, pp. 170–194.

Ratcliff, N. J., K. Hugenberg, E. R. Shriver, & M. J. Bernstein (2011) "The allure of status: High-status targets are privileged in face processing and memory," ***Personality and Social Psychology Bulletin***, 37(8), pp. 1003–1015.

Read, L. E. (1958) "I, Pencil,"The freeman, December ((2016) The freeman, 66(3), Fall, pp. 22–27).

Redhead, G. & R. I. M. Dunbar (2013) "The functions of language: An experimental study,"***Evolutionary Psychology***, 11(4), pp. 845–854.

Reid, V. M., K. Dunn, R. J. Young, J. Amu, T. Donovan, & N. Reissland (2017) "The human fetus preferentially engages with face-like visual stimuli,"***Current Biology***, 27(12), pp. 1825–1828.

Rico-Uribe, L. A., F. F. Caballero, N. Martín-María, M. Cabello, J. L. Ayuso-Mateos, & M. Miret (2018) "Association of loneliness with all-cause mortality: A meta-analysis,"***PloS One***, 13(1), e0190033.

Riedl, K., K. Jensen, J. Call, & M. Tomasello (2012) "No third-party punishment in chimpanzees,"***Proceedings of the National Academy of Sciences***, 109(37), pp. 14824–14829.

Rigdon, M., K. Ishii, M. Watanabe, & S. Kitayama (2009) "Minimal social cues in the dictator game,"***Journal of Economic Psychology***, 30(3), pp. 358–367.

Rilling, J. K., D. A. Gutman, T. R. Zeh, G. Pagnoni, G. S. Berns, & C. D. Kilts (2002) "A neural basis for social cooperation,"***Neuron***, 35(2), pp. 395–405.

Ritov, O., C. J. Völter, N. J. Raihani & J. M. Engelmann (2024) "No evidence for inequity aversion in non-human animals: a meta-analysis of accept/reject paradigms," ***Proceedings of the Royal Society B, Biological Sciences***, 291(2035).

Rotella, A., A. M. Sparks, S. Mishra, & P. Barclay (2021) "No effect of 'watching eyes': An attempted replication and extension investigating individual differences," ***Plos One***, 16(10), e0255531.

Roberts, A. (2011) Evolution: The human story, Dorling Kindersley.

Robertson, C. E., N. Pröllochs, K. Schwarzenegger, P. Pärnamets, J. J. Van Bavel & S. Feuerriegel (2023) "Negativity drives online news consumption," ***Nature Human Behaviour***, 16

Rockenbach, B., & M. Milinski (2006) "The efficient interaction of indirect reciprocity and costly punishment,"***Nature***, 444 (7120), pp. 718–723.

Roth, A. E., V. Prasnikar, M. Okuno-Fujiwara, & S. Zamir (1991) "Bargaining and market behavior in Jerusalem, Ljubljana, Pittsburgh, and Tokyo: An experimental study,"***American Economic Review***, 81(5), pp. 1068–1095.

Rubinstein, A. (1998) Modeling bounded rationality, MIT Press.

Russo, F. (2018) "Loneliness can be toxic,"***Scientific American***, Jan, pp. 64–69.

Sagi, A., & M. L. Hoffman (1976) "Empathetic distress in the newborn,"***Developmental Psychology***, 12(2), pp. 175–176.

Samahita, M. (2017) "Venting and gossiping in conflicts: Verbal expression in ultimatum games,"***Journal of Behavioral and Experimental Economics***, 67, pp. 111–112.

Sánchez-Amaro, A., & F. Rossano, (2021) "Chimpanzees and bonobos use social leverage in an ultimatum game," ***Proceedings of the Royal Society B***, 288(1962), 20211937.

Sanfey, A, G., J. K. Rilling, J. A. Aronson, L. E. Nystrom, & J. D. Cohen (2003) "The neural basis of economic decision-making in the ultimatum game,"***Science***, 300(5626), pp. 1755–1758.

Sapolsky, R. M. (2018). Behave: The biology of humans at our best and worst. Penguin Press.

Scarf, D., K. Imuta, M. Colombo, & H. Hayne, (2012) "Social evaluation or simple association? Simple associations may explain moral reasoning in infants," ***PloS One***, 7(8), e42698.

Schmidt, M. F. H., H. Rakoczy, & M. Tomasello (2012) "Young children enforce social norms selectively depending on the violator's group affiliation,"***Cognition***, 124(3), pp. 325–333.

Schmidt, M. F. H. & M. Tomasello (2012) "Young children enforce social norms,"***Current Directions in Psychological Science***, 21(4), pp. 232–236.

Schlingloff, L., G. Csibra, & D. Tatone (2020) "Do 15-month-old infants prefer helpers? A replication of Hamlin et al.(2007),"***Royal Society Open Science***, 7(4), 191795.

Senju, A., T. Hasegawa, & Y. Tojo (2005) "Does perceived direct gaze boost detection in adults and children with and without autism? The stare-in-the-crowd effect revisited,"***Visual Cognition***, 12(8), pp. 1474–1496.

Senju, A., & M. H. Johnson, (2009) "The eye contact effect: mechanisms and development,"***Trends in Cognitive Sciences***, 13(3), pp. 127–134.

Shalvi, S., M. J. J. Handgraaf, & C. K. W. De Dreu, (2011), "People avoid situations that enable them to deceive others," ***Journal of Experimental Social Psychology***, 47(6), pp. 1196–1106.

Shariff, A. F., & A. Norenzayan (2007) "God is watching you: Priming God concepts increases prosocial behavior in an anonymous economic game," ***Psychological Science***, 18(9), pp. 803–809.

Shariff, A., F., A. K. Willard, T. Andersen, & A. Norenzayan (2016) "Religious priming: A meta-analysis with a focus on prosociality,"***Personality and Social Psychology Review***, 20(1), pp. 27–48.

Shaw, A., S. Choshen-Hillel, & E. M. Caruso (2016) "The development of inequity aversion: Understanding when (and why) people give others the bigger piece of the pie,"***Psychological Science***, 27(10), pp. 1352–1359.

Shaw, A., & K. R. Olson (2012) "Children discard a resource to avoid inequity,"***Journal of Experimental Psychology: General***, 141(2), pp. 382–395.

Shaw, A., & K. Olson (2014) "Fairness as partiality aversion: The development of procedural justice,"***Journal of Experimental Child Psychology***, 119, pp. 40–53.

Shinohara, A., Y. Kanakogi, Y. Okumura, & T. Kobayashi (2020) "How do Children Evaluate the Gossiper of Negative Gossip?,"***Japanese Psychological Research***, 12279.

Shiozawa, Y. (2006) 'rationality', 'procedural rationality', 'substantive rationality', 'situated rationality' and 'goal rationality', in Handbook of Evolutionary Economics, Japan Association for Evolutionary Economics (ed).

Sigelman, C. K., & K. A. Waitzman (1991) "The development of distributive justice orientations; contextual influence on children's resource allocation,"***Child Development***, 62(6), pp. 1367–1378.

Sigmund, K., E. Fehr, & M. A. Nowak (2002) "The economics of fair play," ***Scientific American***, 286(1), pp. 82–87.

Simon, H. A. (1996) Models of my life, Basic books.

Singer, T., B. Seymour, J. P. O'Doherty, H. Kaube, R. J. Dolan, & C. D. Frithe (2004) "Empathy for pain involves the affective but no sensory components of pain,"***Science***, 303(5661), pp. 1157–1162.

Singer T., B. Seymour, J. P. O'Doherty, K. E. Stephan, R. J. Dolan, & C. D. Frith (2006) "Empathic neural responses are modulated by the perceived fairness of others,"***Nature***, 439(7075), pp. 466–469.

Sloane, S., R. Baillargeon, & D. Premack (2012) "Do infants have a sense of fairness? "***Psychological Science***, 23(2), pp. 196–204.

Smith, A. (1767) The theory of moral sentiments: To which is added a dissertation on the origin of languages, 3rd edition, A. Millar, A. Kincaid and J. Bell in Edinburgh; and sold.

Schmidt, M. F., L. P. Butler, J. Heinz & M. Tomasello (2016) "Young children see a single action and infer a social norm: Promiscuous normativity in 3-year-olds," ***Psychological Science***, 27(10), pp. 1360–1370.

Sommerfeld, R. D., H. J. Krambeck, D. Semmann, & M. Milinski (2007) "Gossip as an alternative for direct observation in games of indirect reciprocity,"***Proceedings of the National Academy of Sciences***, 104(44), pp. 17435–17440.

Sommerville, J. A., M. F. Schmidt, J. E. Yun, & M. Burns (2013) "The development of fairness expectations and prosocial behavior in the second year of life,"***Infancy***, 18(1), pp. 40–66.

Sparks, A., & P. Barclay (2013) "Eye images increase generosity, but not for long: The limited effect of a false cue,"Evolution and Human Behavior, 34(5), pp. 317–322.

Stallen, M., F. Rossi, A. Heijne, A. Smidts, C. K. De Dreu, & A. G. Sanfey (2018) "Neurobiological mechanisms of responding to injustice,"***Journal of Neuroscience***, 38(12), pp. 2944–2954.

Steptoe, A., A. Shankar, P. Demakakos, & J. Wardle (2013) "Social isolation, loneliness, and all-cause mortality in older men and women,"***Proceedings of the National Academy of Sciences***, 110(15), pp. 5797–5801.

Stone, V. E., L. Cosmides, J. Tooby, N. Kroll, & R. T. Knight (2002) "Selective impairment of reasoning about social exchange in a patient with bilateral limbic system damage,"***Proceedings of the National Academy of Sciences***, 99(17), pp. 11531–11536.

Strathearn, L., J. Li, P. Fonagy, & P. R. Montague (2008) "What's in a smile? Maternal brain responses to infant facial cues,"Pediatrics, 122(1), pp. 40–51.

Stavrova, O., & D. Ren (2023) "Alone in a Crowd: Is Social Contact Associated with Less Psychological Pain of Loneliness in Everyday Life?,"Journal of Happiness Studies, 24, pp. 1841–1860.

Strelan, P., & J. W. van Prooijen (2013) "Retribution and forgiveness: The healing effects of punishing for just deserts,"European Journal of Social Psychology, 43(6), pp. 544–553.

Strobel, A., J. Zimmermann, A. Schmitz, M. Reuter, S. Lis, S. Windmann, & P. Kirsch (2011) "Beyond revenge: neural and genetic bases of altruistic punishment,"***Neuroimage***, 54(1), pp. 671–680.

Sugiyama, L. S., J. Tooby, & L. Cosmides (2002) "Cross-cultural evidence of cognitive adaptations for social exchange among the Shiwiar of Ecuadorian Amazonia,"***Proceedings of the National Academy of Sciences***, 99(17), pp. 11537–11542.

Surkalim, D. L., M. Luo, R. Eres, K. Gebel, J. van Buskirk, A. Bauman, & D. Ding (2022) "The prevalence of loneliness across 113 countries: systematic review and meta-analysis,"***BMJ***, 376.

Sutter, M., C. Zoller, & D. Glätzle-Rützler (2019) "Economic behavior of children and adolescents – A first survey of experimental economics results,"***European Economic Review***, 111, pp. 98–121.

Svetlova, M., S. Nichols, and C. Brownell. (2010) "Toddlers' prosocial behavior: From instrumental to empathic to altruistic helping," ***Child Development***, 81(6), pp. 1814–1827.

Tabibnia, G., & M. D. Lieberman (2007) "Fairness and cooperation are rewarding: evidence from social cognitive neuroscience,"***Annals of the New York Academy of Sciences***, 1118(1), pp. 90–101.

Tabibnia, G., A. B. Satpute, & M. D, Lieberman (2008) "The sunny side of fairness: preference for fairness activates reward circuitry (and disregarding unfairness activates self-control circuitry),"***Psychological Science***, 19(4), pp. 339–347.

Takahashi, H., M. Kato, M. Matsuura, D. Mobbs, T. Suhara, & Y. Okubo (2009) "When Your Gain Is My Pain and Your Pain Is My Gain: Neural Correlates of Envy and Schadenfreude,"***Science***, 323 (5916), pp. 937–939.

Takai, S., A. Hasegawa, J. Shigematsu, T. Yamamoto (2023) "Do people who highly value happiness tend to ruminate?," ***Current Psychology***, 42, pp. 32443–32455.

Tane, K., & M. Takezawa (2011) "Perception of human face does not induce cooperation in darkness,"***Letters on Evolutionary Behavioral Science***, 2(2), pp. 24–27.

Tania A. Reynolds, T. A., J. K. Maner, R. F. Baumeister (2025) "Bless her heart: Gossip phrased with concern provides advantages in female intrasexual competition," ***Journal of Experimental Social Psychology***, 116, 104670.

Tei, S. & H. Takahashi (2015) "altruistic punishment," in Dictionary of Neuroscience (Nou kagaku ziten). https://doi.org/10.14931/bsd.5897 (in Japanese).

Thaler, R. (2015) Misbehaving: The making of behavioral economics, W. W. Norton & Company.

Thielmann, I., D. W. Heck and B. E. Hilbig (2016) "Anonymity and incentives: an investigation of techniques to reduce socially desirable responding in the Trust Game,"***Judgment and Decision Making***, 11(5), pp. 527–536.

Tomasello, M. (2016) "A natural history of human morality," Harvard University Press.

Tomasello, M. (2019) "Becoming human: A theory of ontogeny," Harvard University Press.

Tomova, L., K. L. Wang, T. Thompson, G. A. Matthews, A. Takahashi, K. M. Tye, & R. Saxe (2020) "Neuroscience: Social isolation evokes craving responses in the human brain,"***Nature Neuroscience***, 23(12), pp. 1597–1605.

Thompson, E. L. C. M. S. Plowright, C. M. Atance, J. S. Caza (2015) "Reasoning and relatedness," Evolution and Human behavior, 36(1), pp. 38–43.

Tricomi, E., A. Rangel, C. F. Camerer, & J. P. O'Doherty (2010) "Neural evidence for inequality-averse social preferences,"***Nature***, 463(7284), pp. 1089–1091.

Trivers, R. L. (1971) "The evolution of reciprocal altruism,"***The Quarterly Review of Biology***, 46(1), pp. 35–57.

Twenge, J. M., R. F. Baumeister, C. N. DeWall, N. J. Ciarocco, & J. M. Bartels (2007) "Social exclusion decreases prosocial behavior,"***Journal of Personality and Social Psychology***, 92(1), pp. 56–66.

Twenge, J. M., R. F. Baumeister, D. M. Tice, & T. S. Stucke (2001) "If you can't join them, beat them: Effects of social exclusion on aggressive behavior,"***Journal of Personality and Social Psychology***, 81(6), pp. 1058–1069.

Twenge, J. M., K. R. Catanese, & R. F. Baumeister (2002) "Social exclusion causes self-defeating behavior,"***Journal of Personality and Social Psychology***, 83(3), pp. 606–615.

Vaish, A., C. M, Kelsey, A.Tripathi, & T. Grossmann, (2017) "Attentiveness to eyes predicts generosity in a reputation-relevant context," ***Evolution and Human Behavior***, 38(6), pp. 729–733.

Valtorta, N. K., M. Kanaan, S. Gilbody, S. Ronzi, & B. Hanratty (2016) "Loneliness and social isolation as risk factors for coronary heart disease and stroke: systematic review and meta-analysis of longitudinal observational studies,"***Heart***, 102(13), pp. 1009–1016.

van Beest, I., & K. D. Williams (2006) "When inclusion costs and ostracism pays, ostracism still hurts,"***Journal of Personality and Social Psychology***, 91(5), pp. 918–928.

van Damme, E., K. G. Binmore, A. E. Roth, L. Samuelson, E. Winter, G. E. Bolton, A. Ockenfels, M. Dufwenberg, G. Kirchsteiger, U. Gneezy, M. G. Kocher, M. Sutter, A. G. Sanfey, H. Kliemt, R. Selten, R. Nagel, & O. H. Azar (2014) "How Werner Güth's ultimatum game shaped our understanding of social behavior," ***Journal of Economic Behavior & Organization***, 108, pp. 292–318.

van den Berg, P., S. Dewitte, & T. Wenseleers (2021) "Uncertainty causes human to use social heuristics and to cooperate more: An experiment among Belgian university students,"***Evolution and Human Behavior***, 42(3), pp. 223–229.

Van Elk, M., D. Matzke, Q. Gronau, M. Guang, J. Vandekerckhove, & E. J. Wagenmakers (2015) "Meta-analyses are no substitute for registered replications: A skeptical perspective on religious priming," ***Frontiers in Psychology***, 1365.

Vanneste, S., J. Verplaetse, A. V. Hiel, & J. Braeckman (2007) "Attention bias toward noncooperative people. A dot probe classification study in cheating detection,"***Evolution and Human Behavior***, 28(4), pp. 272–276.

Vanish, A., M. Carpenter, & M. Tomasello (2010) "Young children selectively avoid helping people with harmful intentions,"***Child Development***, 81(6), pp. 1661–1669.

Vesterlund, L. (2016) "Using experimental methods to understand why and how we give to charity,"in Handbook of experimental economics, (Kagel, J. H., & A. E. Roth (Eds.)), vol.2, pp. 91–151.

von Grünau, M., & C. Anston (1995) "The detection of gaze direction: A stare-in-the-crowd effect,"***Perception***, 24(11), pp. 1297–1313.

Vonasch, A. J., T, Reynolds, B. M. Winegard, & R, F. Baumeister (2018) "Death before dishonor: Incurring costs to protect moral reputation," ***Social Psychological and Personality Science***, 9(5), pp. 604–613.

Vosoughi, S., D. Roy, & S. Aral (2018) "The spread of true and false news online,"***Science***, 359(6380), pp. 1146–1151.

Wang, H., K. Li, & X. Xie (2019) "Individual differences in pain sensitivity predict the experience of unfairness," ***Journal of Health Psychology***, 24(7), pp. 953–963.

Wang, R., Y. Wang, C. Chen, L. Huo, & C. Liu (2023) "How do eye cues affect behaviors? Two meta-analyses,**"** ***Current Psychology***, pp. 1–18.

Wang Y, J. Ge, H. Zhang, H. Wang, X. Xie (2020) "Altruistic behaviors relieve physical pain," ***Proceedings of the National Academy of Sciences***, 17(2), pp. 950–958.

Warneken, F. (2013) "Young children proactively remedy unnoticed accidents," ***Cognition***, 126(1), pp. 101–108.

Warneken, F., and M. Tomasello (2006) "Altruistic helping in human infants and young chimpanzees," ***Science***, 311(5765), pp. 1301–1303.

Warneken, F., and M. Tomasello (2007) "Helping and cooperation at 14 months of age," ***Infancy***, 11(3), pp. 271–294.

Warneken, F., and M. Tomasello (2013) "The emergence of contingent reciprocity in young children," ***Journal of Experimental Child Psychology***, 116(2), 338–350.

Wascher, C. A. F., & T. Bugnyar (2013) "Behavioral responses to inequity in reward distribution and working effort in crows and ravens,"***PLoS ONE***, 8(2): e56885.

Watanabe, M. (2013) "prefrontal area"in Dictionary of Neuroscience (nou kagakuziten). https://doi.org/10.14931/bsd.1657 (in Japanese).

Watanabe, S. (2020) "The animal in you: welcome to the world of comparative cognitive science (Anatano naka no doubutsu tachil Youkoso hikaku ninchi Kagaku no Sekai he)," Kyoiku Hyoron sha (in Japanese).

Watanabe, S., & K. Ono (1986) "An experimental analysis of"empathic"response: Effects of pain reactions of pigeon upon other pigeon's operant behavior,"***Behavioural Processes***, 13(3), pp. 269–277.

Wiessner, P. (2005) "Norm enforcement among the Ju/'hoansi bushman,"***Human Nature***, 16(2), pp. 115–145.

Wiessner, P. (2014) "Embers of society: Firelight talk among the Ju/'hoansi Bushmen," ***Proceedings of the National Academy of Sciences***, 111(39), pp14027–14035.

Wilkes, R. (2002) Scandal: A Scurrilous History of Gossip, Atlantic.

Wilkinson, R., & K. Pickett, (2019) "The inner level: How more equal societies reduce stress, restore sanity and improve everyone's well-being,"Penguin.

Williams, K. D., C. K. Cheung, & W. Choi (2000) "Cyberostracism: effects of being ignored over the Internet,"***Journal of Personality and Social Psychology***, 79(5), pp. 748–762.

Williams, K. D., & S. A. Nida, (2011) "Ostracism: Consequences and coping,"***Current Directions in Psychological Science***, 20(2), pp. 71–75.

Wilson, D. S. (2007). Evolution for everyone: How Darwin's theory can change the way we think about our lives. Delta.

Xiao, E., & D. Houser (2005) "Emotion expression in human punishment behavior,"***Proceedings of the National Academy of Sciences***, 102(20), pp. 7398–7401.

Xu, X., X. Zuo, X. Wang, & S. Han (2009) "Do you feel my pain? Racial group membership modulates empathic neural responses,"***Journal of Neuroscience***, 29(26), pp. 8525–8529.

Xue, J. & Y. Ohtsubo (2020) "Altruistic Preferences: in the dictator game replication of Android Miller (2002) in Japan,"***Letters on Evolutionary Behavioral Science***, 11(1), pp. 17–21.

Yamagishi, T. (1986) "The provision of a sanctioning system as a public good," Journal of Personality and Social Psychology, 51(1), pp. 110–116.

Yamagishi, T., Y. Li, A. S. R. Fermin, R. Kanai, H. Takagishi, Y. Matsumoto, T. Kiyonari, & M. Sakagami, (2017) "Behavioural differences and neural substrates of altruistic and spiteful punishment," ***Scientific Reports***, 7(1), pp. 1–8.

Yamagishi, T., & N. Mifune (2008) "Does shared group membership promote altruism? Fear greed and reputation,"***Rationality and Society***, 20, pp. 5–30.

Yamagishi, T., Y. Horita, H. Takagishi, M. Shinada, S. Tanida, & K. S. Cook (2009) "The private rejection of unfair offers and emotional commitment," ***Proceedings of the National Academy of Sciences***, 106(28), pp. 11520–11523.

Yamagishi, T., Y. Horita, N. Mifune, H. Hashimoto, Y. Li, M. Shinada, A. Miura, K. Inukai, H. Takagishi, & D. Simunovic (2012) "Rejection of unfair offers in the ultimatum game is no evidence of strong reciprocity,"***Proceedings of the National Academy of Sciences***, 109(50), pp. 20364–20368.

Yamagishi, T., Y. Li, A. S. R. Fermin, R. Kanai, H. Takagishi, Y. Matsumoto, T. Kiyonari, & M. Sakagami (2017) "Behavioral differences and neural substrates of altruistic and spiteful punishment,"***Scientific Reports***, 7(1), 14654.

Yamagishi, T., S. Terai, T. Kiyonari, N. Mifune, & S. Kanazawa (2007) "The social exchange heuristic: Managing errors in social exchange,"***Rationality and Society***, 19(3), pp. 259–291.

Yamamori, T., K. Kato, T. Kawagoe, & A. Matsui (2008) "Voice matters in a dictator game,"***Experimental Economics***, 11, pp. 336–343.

Yamamori, T., K. Kato, & A. Matsui (2010) "When you ask Zeus a favor: The third party's voice in a dictator game,"***Japanese Economic Review***, 61(2), pp. 145–158.

Yasue, M., A. Nakagami, K. Nakagaki, N. Ichinohe, & N. Kawai (2018) "Inequity aversion is observed in common marmosets but not in marmoset models of autism induced by prenatal exposure to valproic acid,"***Behavioural Brain Research***, 343, pp. 36–40.

Yudkin, D. A., J. J. Van Bavel, & M. Rhodes (2020) "Young children police group members at personal cost,"***Journal of Experimental Psychology***, 149(1), pp. 189–191.

Zadro, L., K. D. Williams, & R. Richardson (2004) "How low can you go? Ostracism by a computer is sufficient to lower self-reported levels of belonging, control, self-esteem, and meaningful existence,"***Journal of Experimental Social Psychology***, 40(4), pp. 560–567.

Zak, P. J. (2012) The moral molecule: The source of love and prosperity, Dutton.

Zhang, L., & A. Ortmann (2014) "The effects of the take-option in dictator-game experiments: a comment on Engel's (2011) meta-study,"***Experimental Economics***, 17(3), pp. 414–420.

Zhang, L. S., Ibrahim, N., & Basu, S. (2024) Whispered words and organizational dynamics: The nuanced evaluation of gossipers' personality and its effect on workplace advice seeking. ***Journal of Experimental Social Psychology***, 115, 104643.

Zhong, C. B., V. K. Bohns, & F. Gino (2010) "Good lamps are the best police: Darkness increases dishonesty and self-interested behavior,"***Psychological Science***, 21(3), pp. 311–314.

Zsidó, A. N., C. Bali, F. Kocsor, & M/ C. Hout (2022). Task-irrelevant threatening information is harder to ignore than other valences. Emotion (Washington, D.C.), https://doi.org/10.1037/emo0001189

⟨Max Planck Institute website: interview with Gute.⟩http://www.econ.mpg.de/english/research/ESI/gueth_interview.php (Date accessed 2019.12.17)

⟨TED Talk by de Waal.⟩ https://www.ted.com/talks/frans_de_waal_moral_behavior_in_animals. (Date accessed 2020.2.19)

⟨Kyoto University HP: Born to love superheroes⟩ https://www.kyoto-u.ac.jp/en/research-news/2017-04-06 (Data accessed 2023.6.11)

⟨Wall street journal HP: "The Civilizing Effect Of Market Economics"⟩ https://www.wsj.com/articles/SB1011885700476777840(2002.1.24) (Date accessed 2019.11.12)

The manufacturer's authorised representative in the EU is Springer Nature Customer Service Centre GmbH, Europaplatz 3, 69115 Heidelberg, Germany. If you have any concerns regarding our products, please contact ProductSafety@springernature.com

Printed and bound by CPI Group (UK) Ltd, Croydon, CR0 4YY
30/09/2026
02225308-0016